THE FUTURE OF WORK AND HEALTH

THE FUTURE OF WORK AND HEALTH

The Institute for Alternative Futures

CLEMENT BEZOLD

RICK J. CARLSON

JONATHAN C. PECK

Auburn House Publishing Company
Dover, Massachusetts • London

Library of Congress Cataloging in Publication Data

Bezold, Clement.
The future of work and health.

Includes index.
1. Industrial hygiene—United States. 2. Labor and laboring classes—Health and hygiene—United States. 3. Population forecasting—United States. 4. United States—Occupations—Forecasting. I. Carlson, Rick J. II. Peck, Jonathan C. III. Institute for Alternative Futures. IV. Title.
HD7654.B49 1986 331.25 85-18627
ISBN 0-86569-088-X

FOREWORD

Our understanding of the trends that affect the interplay of work and health is greatly enhanced with *The Future of Work and Health*. As a cornerstone of the national initiative in this area, this book will complement our efforts to guide the development of a new generation of worksite health promotion. Our knowledge of the synergism of environmental and personal risk factors has brought us into an age where health promotion is not a separate endeavor conducted at the worksite but rather an intrinsic component of occupational health. We are no longer asking if the worksite will be used to promote the enhancement of personal health, but, rather, what are the most effective mechanisms for preventing health problems associated with work. *The Future of Work and Health* contributes importantly to our understanding of how work, workers, and the health care system are changing. It is a valuable resource for decision makers in the worksite and public policy arenas whose policies and strategies can be designed to capitalize on the changes and trends presented in these pages. Without the ability to look ahead, our planning will be tied to perceptions of the past rather than to opportunities of the future.

For nearly a decade we have watched the workplace serve as an appropriate and effective location for health promotion activities. Worksite programs have succeeded in supporting healthy behaviors, detecting risk factors for illness and early signs of disease, and aiding with the adherence to medical regimens. Since the sponsorship by the Office of Disease Prevention and Health Promotion of the first National Conference on Health Promotion in Occupational Settings in 1979, the Public Health Service has pursued ways to support the implementation of worksite programs to promote health. We have also learned through extensive research how work affects health, and we have witnessed the development of more sophisticated mechanisms to prevent occupational hazards. Indeed, 1985 marks the turning point toward *prevention* at the workplace with the First National Symposium on the Prevention of Leading Work-Related Disease and Injuries sponsored by the National Institute on Occupational Safety & Health.

The Public Health Service is no stranger to planning for the future. Virtually all contemporary health promotion and disease prevention policies and programs in the United States have been greatly influenced by the publication entitled, *Promoting Health/Preventing Disease: Objectives for the Nation*, published in 1980 by the Public Health Service. This document delineates 227 objectives which, if achieved, will substantially improve the health of Americans by the year 1990.

The time is right to focus our attention more sharply on the workplace of the future. *The Future of Work and Health*'s comprehensive presentation of changing trends in work and health allows us to consider health promotion and disease prevention in the light of the special and newly evolving characteristics of work. These trends help us to anticipate the characteristics of work and the workforce of the future so that medical, disability, and retirement benefits; work schedules; education opportunities; occupational health and safety programs; communication strategies; peer support systems; and many other aspects of work can be designed to help individuals collectively achieve their highest level of health through prevention. Prospective health strategies offer an exciting potential for improving the health of workers, retirees, and the families of workers—opportunities for health that affect all of us. This book represents the beginning of our efforts to tap the potential of the future of work and health. We commend its authors and the many people who contributed their time to making this document comprehensive and thought-provoking. We hope that all who read it join in the commitment to making the future healthier.

J. Michael McGinnis, M.D.

Deputy Assistant Secretary for Health and
Director, Office of Disease Prevention and Health Promotion
Public Health Service

J. Donald Millar, M.D.

Assistant Surgeon General and
Director, National Institute for Occupational Safety and Health
Centers for Disease Control
U.S. Department of Health and Human Services

PREFACE

The purpose of the project on the future of work and health was simple: to identify the most important characteristics of work and workplaces over the next twenty-five years, particularly in relation to health issues. The Office of Disease Prevention and Health Promotion (ODPHP) and the National Institute for Occupational Safety and Health (NIOSH) of the Department of Health and Human Services sponsored the project because the U.S. Public Health Service shares an interest in the responsibility to challenge employers and employees to create healthy work and healthy workplaces. This project was a fundamental and necessary step in achieving that mission.

The design and implementation of programs to prevent injuries, disabilities and illnesses on the job and to enhance health ought to be aided by an effort to understand the future of work and the workplace. Hence, this project focused on the context within which future work-related health programs will take place. Yet, just as work will change in the years ahead, so will health and health care. Key trends shaping health care will shape health promotion programs that will affect workers and workplaces in the future.

In the following chapters, a variety of trends which are shaping the future of work and health will be discussed. We shall not undertake an exhaustive study of the future of work and health; this would be impossible in this limited space because work and health are too extensive, affecting every facet of our lives. We shall, however, identify the most important trends to monitor—a necessary step if we are to understand change.

Chapter 1 provides a snapshot of key changes in work and health expected for the early part of the next century. It also introduces future thinking, and puts forth the most important questions raised when this project was initiated. Chapter 2 presents some of the basic assumptions about the U.S. population and workforce on which this work was based, as well as some important demographic trends. The basic data in Chapter 2 will be used to analyze the impacts of the trends identified in Chapters 3 and 4. The key trends affecting work, particularly in the areas of the economy's fundamental nature, values, technology, and changes in the workplace itself, are reviewed in Chapter 3. Chapter 4 presents the key trends shaping health and health care, particularly in the areas of health care delivery, technologies, and our conceptions of health care. Chapter 5 draws implications from the earlier chapters and the October 1984 workshop which reviewed this report; Chapter 6 provides a summary of all of the chapters and an image of promotion of health at the dawn of the twenty-first century. The bibliography is divided into sections on the future of work and the future of health. Included in the section on the future of work are the sources referred to in Chapter 2 (on population and demographics) as well as those used in Chapter 3 (on work).

As you read this book, keep in mind that change, especially in this era, is more often dialectical than evolutionary. The trends are identified separately, as if each were evolving independently. Yet they will interact with each other, often in contradictory ways. Also, these trends represent our thinking in 1984 and 1985 about the most important forces shaping work and health. There will be surprises. The uncertainty that emerges from the interactive nature of the trends and from unforeseen influences does not diminish the importance of understanding each of the key forces identified here.

A critical task of this project is to make the uncertainty of the future more visible and more comprehensible, so that decisions in the next few years can be made as wisely as possible. Understanding key trends and the dialectical nature of change enhances that capacity for wiser decision-making.

Based on our expertise as well as a review of the literature and comments from leading experts, this book explores the often conflicting assumptions about the future of work and health. The implications of trends and forecasts are then considered, particularly the implications for promotion of health in the workplace. Most of the trend literature focuses on what has been happening during the last few years, while some explores the next few years. We will use both, often extending out to the first part of the twenty-first century.

It is important for each of us as individuals to consider the future. Reaching out to the next century forces us to confront the uncertainty inherent in the future and to enhance our capacity to deal with this uncertainty. Indeed, as Donald Michael has argued, dealing with this uncertainty, including the recognition that there is much that is beyond our control, is among the "new competencies" required of decision-makers.

ACKNOWLEDGMENTS

This book is the result of a report prepared for the Office of Disease Prevention and Health Promotion and the National Institute of Occupational Safety and Health by the Institute for Alternative Futures in conjunction with the Washington Business Group on Health. It was developed with assistance from a large number of individuals and organizations.

The principal authors are Clement Bezold, Executive Director of the Institute for Alternative Futures, Jonathan C. Peck, Associate Director, and Rick Carlson. On the Institute for Alternative Futures staff, J. Richard Scarce, Katherine Scott, and Brenda Cavanaugh were directly involved. For OHPDP, Anne Kiefhaber, Director of National Worksite Health Promotion Initiative, was the project officer and the originator of the concept for the report and the larger project of which this is a part. Cheryl Damberg, Research Fellow for OHPDP, was also a contributor.

A critical step in developing this material was the convening of an advisory panel of experts on health and work in October of 1984. These individuals are listed below. They not only provided comments at the session on the initial draft but in many cases also critiqued a subsequent version of the report. Three others who were not in attendance provided crucial data or advice or reviewed part or all of the report. These were Trevor Hancock, Associate Medical Officer for Health, City of Toronto; John Naisbitt, author of *Megatrends*; and Kenneth Pelletier, author of *Unhealthy People, Unhealthy Places*.

ADVISORY PANEL

Louis Beliczky
Director of Industrial Hygiene
URW International Union

Arnold Brown
President
Weiner, Edrich & Brown, Inc.

Gil Collings, M.D.
Retired Medical Director
New York Telephone

Cheryl Damberg
Research Fellow
ODPHP

James F. Fries, M.D.
Associate Professor
Stanford University—
School of Medicine

Willis Goldbeck
President
Washington Business Group
on Health

Robert Graham, M.D.
Vice President and
Chief Medical Officer
Equitable Life Assurance Company

Doug Greene
Chairman
New Hope Communications

Anne Kiefhaber
Project Director
Worksite Health Promotion
Initiative
Department of Health &
Human Services

Hank Koehn
Vice President
Futures Research Division H10-1
Security Pacific National Bank

Sar Levitan, Ph.D.
Director
Center for Social Policy Studies
George Washington University

Stanley J. Matek, Ph.D.
Past President
American Public Health Association

J. Michael McGinnis, M.D.
Director
Office of Disease Prevention and
Health Promotion
Deputy Assistant Secretary
Department of Health &
Human Services

Donald N. Michael, Ph.D.
Professor Emeritus of Planning and Public Policy
University of Michigan

J. Donald Millar, M.D.
Director
National Institute of Occupational Safety and Health
Center for Disease Control

Melvin Myers
Director
Office of Program Planning & Evaluation
National Institute of Occupational Safety and Health
Center for Disease Control

Phillip Polakoff, M.D.
Principal
Integrated Health Management Associates

Elsa A. Porter
Consultant

Marie Spengler
President
Assisted Intelligence Design, Inc.

CONTENTS

LIST OF FIGURES

LIST OF TABLES

THE FUTURE OF WORK AND HEALTH

Chapter 1

A SNAPSHOT OF THE FUTURE OF WORK AND HEALTH

By the first part of the twenty-first century, as we work and strive to promote our health, we will have experienced a wide variety of changes. Some of the following trends have been predicted in the work and health arenas:

- Many of the information and service jobs, as well as those in manufacturing, have been eliminated due to automation and artificial intelligence. The middle ranks of large and small organizations have thinned, leaving—out of a potential workforce (civilian, non-institutionalized, 20 years old and over) of about 200 million—a workforce of 110 million. Ten million of the 110 million workers share a job, and the average work-week is down to 32 hours.
- A large number of management tasks, as well as whole jobs, have been eliminated, and those who still work are dramatically more productive. The development of expert systems (artificial intelligence in computer software) has allowed machines to closely approximate human thought processes.
- In manufacturing, robots and computer-assisted design and computer-assisted manufacturing (CAD/CAM) have eliminated a significant percentage of jobs.
- More people are over 65, but the elderly and everyone else suffer from fewer disabling conditions over the course of their lives, thus compressing morbidity until

the very end of life. This has resulted in large part from greater attention to the health fundamentals of diet, exercise, and stress management, as well as from the reduction or avoidance of alcohol, cigarettes, and other harmful drugs.

- The medical care system is still extensive and costly. Yet not only do a wide variety of physicians, nurses, and alternative practitioners provide care, but individuals routinely care for themselves using a variety of "self-care" devices (such as the "hospital-on-the-wrist," a miniaturized wristwatch-size device that monitors, diagnoses, and treats health conditions) as well as their own enhanced "body wisdom."
- Because of problems resulting from unrecognized toxic exposures in the 1990's, the pressures on occupational health and safety programs have intensified and toxins are under much tighter control. Workers' genetic sensitivity to specific toxins is known, and while genetic engineering can be used to desensitize workers, few employers use it.
- The informal economy, the cash portion of which was about 15 percent of the formal GNP in the early 1980's, has expanded dramatically. This was driven by the growing scarcity of formal jobs and by value shifts that put greater emphasis on personal growth and community service. This transition in values shifted the conception of work and purpose, bestowing greater meaning on nonpaid work. It has also enabled communities, families, and churches to reappropriate the care that was lost as welfare services became monetized during the twentieth century.
- Of the 14 percent of the population that is black in 2010, many are still members of the underclass, which also includes Hispanic immigrants and whites. The social unrest from this group would be greater were it not for the growth of the informal economy and community self-help.

Given their assumptions, each of these statements is a likely forecast. We will explore these forecasts, as well as a wide range of others, and examine the often-conflicting assumptions about the future of work and health. Based on a review of the literature and comments from leading experts, we will identify key trends. At various points we will use these trends to speculate about the next century. Some possible implications of these trends and forecasts, particularly implications for the workplace, will then be posited.

Thinking About the Future of Work

A primary objective here is to identify the trends and shaping forces that will be important in affecting the nature of work and health in the first part of the twenty-first century. A review of the literature easily identifies a host of issues, outlined in the following sections. First some comments are in order about the hesitancy of our imagination when confronted by such a task. Experience shows that the future will bring surprises.

In pondering the nature of work in the early part of the twenty-first century, we might take heart in Isaac Asimov's opening remarks to a 1979 meeting on work in the twenty-first century:[1]

> *It may be that the twenty-first century will be the great age of creativity in which finally machines will do the humdrum work of humanity. The computers will keep the work going, and human beings will be free at last to do things that only human beings can do—to create . . . .*
>
> *The work of the future will be creation, done by each in his own fashion . . . . It is difficult to work out the details of such a world because we continue, even against our will, to think in the terms of the world as it now exists. If this were 1800 and I tried to describe the world of 1979, and if I said that people would be moving around at unprecedented rates and constantly in communication with each other over huge distances, the question might be, "How does one develop horses that run that fast?" Or "What kind of megaphone do you use to enable you to shout loud enough for one to hear you miles away?"*

. . . . It would be better if instead of one person working on the problem, we have groups in which one will advance one clever facet and another will advance another clever facet and out of all of them put together we then may begin to get a glimpse of a twenty-first century. This much I know: it will be as far advanced beyond ourselves as we are beyond the Middle Ages.

Remember, though, that all of this presents a deliberate and voluntary choice on the part of humanity . . . . I hope we all remember that whenever we look forward into the future, we are looking forward to something about which we have a choice. That is for our benefit and our good—if we choose wisely.

Arnold Brown and Edith Weiner see significant, although not necessarily radical, changes over the next twenty years. They remind us of the dramatic change that has occurred over the last twenty:[2]

If Ms. Executive, a well-schooled senior manager, were to wake up one morning and suddenly find herself, like Rip Van Winkle, 20 years hence, she would probably find it less shocking than if she were to wake up 20 years back. Placed in the past, she would be appalled by the outright discrimination she would face at work, by the very limited expectations her husband and family would have of her, by the primitive technology she would have available to calculate and type, by the refusal of the manufacturers of goods she purchased to accept returns of faulty or malperforming products, by the smokestacks in the city belching out black soot, by the lack of readily available means to detect or cure breast cancer, by the limited media available to her to view and read, by the comments she would get from neighbors if her floors were not spic and span all of the time, by the widespread belief that we could not reach the moon.

Our task in Chapter 1 of *The Future of Work and Health* is to imagine, not the full spectrum of the twenty-first century, but only its first decade. A diverse set of trends are identified in Chapters 2, 3, and 4. This chapter highlights those trends and poses key questions about the forecasts presented here.

Work and the Workplace

As a society we are in a time of significant change. As John Naisbitt observes, "We are living in *the time of the parenthesis*, the time between eras. It is as though we have bracketed off the present from both the past and the future, for we are neither here nor there."[3]

These tides of change powerfully influence our perceptions about work and its place in our society. Only in the last years has public attention been drawn to such issues; although, of course, commentators like Daniel Bell, Donald Michael, James O'Toole, Sar Levitan, and Alvin Toffler have been pointing out the change to come for longer than that.[4-8] Still, the impact of the computer and other new communications technologies, the rise of foreign competition—not just in computers and electronics, but in our basic industries such as steel and autos—and the persistence of high unemployment rates (relative to the 1950's and 1960's) are just some of the issues with which the American public is becoming familiar.

The Change in Perception

We are in a period of great transition—including a transition in our perceptions of work and the workplace. For some workers and employers, perceptions of work and the workplace are changing, and the momentum is growing. These changes are occurring in at least four ways: the first is our changing economy. The explanatory power of basic economic concepts and measures; the ability of these same measures to encompass quality of life; the size and importance of the "informal" economy; the impact of new technologies on job formation, job retention, and workplace environments; changing patterns of compensation, such as merit pay; and expected rates of economic growth—all are just a few of the key issues.

The second change involves the employer's perception of the worker. Employers perceive workers as something more than interchangeable parts in the production machine. But this change is accompanied by the difficulty of grasping what an "information" worker is—not merely somebody who comes to work and gives "8 for 8"—and how such workers should be treated and managed. Employees are being given more authority in the management of the workplace as well, and there are fewer layers of supervisors between workers and managers. These changes accompany the acknowledgment—after centuries of work—that many employees seek more from their jobs than just an income, along with a focus on the relationship between new forms of management and productivity.

The third change is the workers' perceptions of work. For increasing numbers of workers, issues coming to the fore include productivity and its relationship to quality; what rewards, incentives, and recognition workers seek and need; the degree to which participation in work-related decisions is important; perceptions by employees of the job and the workplace as settings in which many of the new "expressive" values can be experienced and sought; and very fundamentally, any change in the value attached to work for its own sake, particularly given the possibility that many of the new technologies for use in the workplace may render unnecessary much of the work now done.

The fourth change is in perceptions of the workplace. New communications technologies may enable more work to be done from the home or in worksites with smaller, more human scales. Increasingly, workers are pursuing changes in the workplace to promote health and well-being, along with other self-enhancement opportunities. The workplace may be evolving into a setting in which a larger variety of human needs are met and human services provided.

The Key Questions

What are the definitive questions—those questions that, if answerable, would offer a fairly clear picture of the future? Exploration of such questions will allow a clearer understanding of how shifts in perception may result in real institutional change. There are at least eight such key questions:

1. What will be the impact of new (and newer) technologies, particularly information and communications technologies, on job formation; job displacement; training, retraining, and educational programs; workplace structure and relationships; people's perception of their jobs; and purchasing power as a function of income from work?
2. Will there be any major private sector and public policy shifts that dramatically alter the workplace landscape? Shifts of this magnitude might unfold in one or more of the following areas:
   a. New public policies might combine aspects of all these worksite changes and other discrete initiatives.
   b. Programs of any sort to fix or stabilize unemployment levels at or near "full employment" (whatever standard that might be), whether involving some form of "guaranteed income," reduced workweek, or public works and/or "reindustrialization" variants that have this effect, or some combination of these.
   c. Major new training and retraining programs, of significantly greater magnitude than currently provided, that focus on jobs in new workplaces spawned by new technologies.
   d. "Revitalization" or "reindustrialization" programs representing significant rearrangements in the relationships between and among labor, management, and government, along with substantially altered incentive and/or subsidy programs.
   e. Changes in attitudes about, and development of mechanisms for, distributing income not tied to jobs and work, which both cushion the displacement of persons out of the full-time workforce and/or offer real alternatives to employment. Will these new programs be any less "disempowering" than traditional welfare programs?

3. Will the means be developed to integrate informal economic activity into the formal economy, among other things, to create a value structure and reward system for that type of activity that will be perceived as work in much the same way as work is perceived and rewarded in today's formal economy?
4. What will be the continuing impact of foreign competition on both U.S. basic industries and new technology markets—for example, intensification of steel and auto competition, and competition from the fifth-generation computers developed in Japan?
5. What will be the impact of changes in the expectations and market behavior of consumers? The public's interest in "health and health enhancement" arose, as do many consumer demands, needs, and interests, in large part from the grassroots. Hence, what changes in consumer demand and needs might occur which would necessarily affect "supply" and in turn the types of jobs in the future?
6. Will society as a whole, those who shape the workplaces of the future, the type of work to be done, and its rewards, continue to change its perceptions of employees as a form of human capital? For example, will concepts of lifetime employment emerge in the United States? (Control Data recently announced its intention to formulate some type of lifetime employment program.)
7. What will be the scale of the workplace? That is, will the new types of jobs allow more human scale for the workplace, which, among other things, would allow for greater outlets for many of the new "expressive" values?
8. And, last but not least, will the value shift described in Chapter 4—the shift toward more expressive values—persist, intensify, take on new forms, diminish, or evolve in some other way?

Health and Health Care

The health field is in the midst of turbulent change. The last ten years have seen a wave of revolutionary ideas and concepts, chief among them the holistic health movement, the renewed interest in self-help and self-care, and the rise of the promotion of health and wellness ideas and programs. Not surprisingly, these ideas have run in tandem with (and some would say were the precursors of) the imposition of limits to growth in the medical sector, the defrocking of some of the priesthoods of modern medicine, the enormous pressures for cost containment, the inexorable tide of consumer disaffection with the medical care system, the widespread restructuring of the health industry, the prospects of more profound research breakthroughs, and, in particular, growth of the for-profit sector in health care. It is no overstatement to say that these new ideas about health, combined with the stirrings of organizational and institutional change in the health system, presage a decade, beginning about now, in which the health care system will change more than it has changed in the last fifty years. This widespread, indeed profound, change is also a reflection of changing perceptions about health.

The Change in Perceptions

As in the case of change in the workplace and in views of work, a small number of shifts in perception lie at the roots of change in health care. The role of medicine (or medical care as a therapeutic system) and of the healing professions is changing. The limits of medicine—argued forcefully by Ivan Illich in *Medical Nemesis,*[9] Rick Carlson in *The End of Medicine,*[10] and more recently by Paul Starr in *The Social Transformation of American Medicine*[11]—are slowly being recognized, and that recognition is reshaping the delivery of health care. Today, few knowledgeable observers of the health scene argue for larger social investments in the medical care system *if* the objective is improved health for our population.

This changing perception of medicine's role, especially when contrasted with opportunities for improving health by other means, is a major factor in the current climate in medicine that is fostering cost management. As health care organizations seek to become businesslike and more competitive, they are imposing tests of cost effectiveness, limits on capital formation and expenditure, and a restructuring of the system, particularly the hospital sector.

The role of the buyers of medical care and health services and of individuals is also changing. Individuals, groups and organizations have begun to scrutinize medicine's products and to demand more in terms of scope (prevention, promotion of health, and so forth), quality, and effectiveness. Purchasers also have become more sophisticated, seeking cost-effective options in health care. For decades the individual has behaved, and been treated, like a passive consumer of services. The participatory ethic and the holistic health, promotion of health, wellness, and self-help movements have combined to create greater self-awareness by many of the role they as individuals can play, not only in the enhancement of health, but in the prevention, care, and management of disease as well.

As the individual's role is being redefined, the role of the mind of that individual is emerging as a potentially powerful mediator of the body's reactions to health threats. One of the most active areas of bioscience today is the further exploration of mind-body dynamics, including the effects of stress on the onset of disease and the enhancement of health, and the role of emotions. An indication of the fast rate of change in the health field—a *change in perception*—is the observation that most of the topics now seriously explored under the rubric "mind-body" interactions were dismissed as rank quackery by medicine as little as ten years ago.

The last shifting perception is with respect to the role of the institutions of health: hospitals, medical clinics, suppliers, payers, and educational programs. Because of all the other changing perceptions, the institutions of health will necessarily undergo radical change in the years between now and 2010, perhaps to the degree that they would be unrecognizable today.

The Key Questions

As in the case of change in the nature of work and the workplace, there are a few key questions about change in the health field that, if answered, would provide a much clearer picture of the future:

1. Will evidence for the compression of morbidity, set forth by James Fries[12] and others, mount to the degree that substantially more resources for disease prevention and health promotion will be allocated? This outcome, of course, has profound implications for workforce issues in the future.
2. Will the impacts of the new health promotion and wellness programs on health status indicators become more evident and demonstrably health-producing, so that more powerful claims to limited resources can be asserted?
3. Similarly, will the new medical technologies, including genetic engineering, organ transplantation, and expert medical computer systems, fuel the further expansion of highly professionalized, "scientific," reductionist medicine? Or will these new technologies be applied so that individuals, in their pursuit of health, are empowered to reduce their dependence on professionals and technical fixes through a more comprehensive understanding of health as well as direct access to sophisticated care?
4. How central is the role of the health profession in determining the distribution of resources our society has set aside for its health? Will the resistance to change already plainly visible in the health field (and not just for economic reasons) be sufficiently strong to halt or delay change? As an aspect of this resistance, will the therapeutic community

successfully "medicalize" the human problems of stress and coping, chemical and alcohol abuse, "environmental" illnesses and maladaptions, as well as the "technology" of high individual productivity and personal performance, so that its grip is at least maintained or even expanded? Or will new institutions arise to provide the means to deal with these new goals that allow fuller expression of individual differences or more latitude for self-help?

5. Will emerging concepts of organizational and community health, the "healthy corporation," and so forth, take hold, along with other health and biology-rooted metaphors, so that our organizational, community, and social lives may be better understood in terms of health? To what extent will the proposition that a healthy individual makes a more productive worker be applied in the converse: that a healthy job makes a healthy individual?

6. To what extent will any of the gains in health from the changes taking place in the health field be undercut, compromised, or even nullified, by the potential of a steady harvest of toxins in our enviroment, the yield of decades of unwitting environmental despoilation?

7. What will be the impact of the new "competitiveness," aggressive marketing, and the "restructuring" of the health care industry into fewer, larger, more "businesslike" organizations on (a) self-help, (b) growth rates in the industry, (c) innovation, (d) health promotion and (e) employer responsibilities for health?

8. What new diseases will be generated from changes in the workplace? For example, will the ergonomic problems of video display terminals become a significant issue?

The task of the following chapters will be to identify the trends that will determine the answer to these questions and to serve as a starting point for discussion. Chapter 2 discusses demographic trends in the population and the workforce. The major trends shaping the nature of work and the workforce and health care are reviewed in detail in Chapters

3 and 4, respectively. Chapter 5 will discuss the implications of future programs for promotion of health in the workplace. The conclusions and a concise summation of the future of health and work will be presented in Chapter 6.

Notes

1. Isaac Asimov, "The Permanent Dark Age: Can We Avoid It?" *Working in the Twenty-First Century*, C. Stewart Sheppard and Donald C. Carroll, eds. (New York: John Wiley & Sons, 1980), pp. 1-11.
2. Arnold Brown and Edith Weiner, *Supermanaging* (New York: McGraw-Hill, 1984), p.14.
3. John Naisbitt, *Megatrends* (New York: Warner, 1982), p. 279.
4. Daniel Bell, *The Coming of Post-Industrial Society: A Venture in Social Forecasting* (New York: Basic Books,1973).
5. Donald N. Michael, *Cybernation: The Silent Conquest* (Santa Barbara, Calif.: Center for the Study of Democratic Institutions, 1962).
6. James O'Toole, *Making America Work: Productivity and Responsibility* (New York: Continuum, 1981).
7. Sar A. Levitan and Clifford M. Johnson, *Second Thoughts on Work* (Kalamazoo, Mich.: W.E. Upjohn Institute for Employment Research, 1982).
8. Alvin Toffler, *The Adaptive Corporation* (New York: McGraw-Hill, 1985).
9. Ivan Illich, *Medical Nemesis: The Expropriation of Health* (New York: Bantam Books, 1976).
10. Rick J. Carlson, *The End of Medicine* (New York: John Wiley, 1975).
11. Paul Starr, *The Social Transformation of American Medicine* (New York: Basic Books, 1982).
12. James F. Fries, "Aging, Natural Death, and the Compression of Morbidity," *New England Journal of Medicine* 303 (1980).

Chapter 2

POPULATION AND WORKFORCE FORECASTS

In this chapter, we set forth some basic data on the size and nature of the population, including major demographic trends relevant to considering the future of work, and basic, largely extrapolative forecasts for the workforce through 2010 used by the Bureau of the Census and the Social Security Administration. Divergent trends which might alter these basic forecasts are considered elsewhere in this book.

Population

The Census Bureau's middle series projections forecast a total U.S. population of 283 million persons by 2010, up from 236 million in 1984. The projection assumes net immigration to be 450,000 per year, slow gains in life expectancy (to age 81 by 2080), and fertility of 1.9 births per woman. Table 2-1 provides the results to 2080 for the middle series, as well as the Census Bureau's high and low estimates. Major demographic uncertainties that might alter that 283 million figure or affect work within that range include mortality and fertility patterns, immigration trends, the racial and ethnic mix, and the geographic distribution of the population.

Worker Demographics

A variety of demographic factors will affect the supply and type of workers over the next twenty-five years, including the basic age distribution, family structure, mortality and morbidity patterns, immigration, the role of minorities, and geographic shifts.

More Women, More Older Workers

Figure 2-1 graphically illustrates probable future changes in the age distribution of the United States, given recent historical fertility and mortality trends. As we move toward 2010, the numbers of women and the elderly (especially elderly women) will increase, and there is reason to believe that older citizens will seek work as they never have before, especially if morbidity patterns continue shifting toward the very late years of life (this is due to the "squaring of the survival curve" trend, reviewed in Chapter 4). However, it would be a mistake to assume that simple numbers alone are an adequate guide. Thus far, even with a larger number of older, and presumably healthier, people, there has been little, if any, measurable pressure on the labor market.

Further, the demands women will place on the labor market may be focused more on the quality of work—the form, style and conditions of their employment—rather than merely on finding jobs. The current trends suggest that women (and men) are seeking to balance career with family and that the pressure for more flexible working arrangements will therefore grow, including demands for company-sponsored day-care, part-time work, and childbirth leave for both parents.[1,2]

As noted in Table 2-2, in 1980 there were about 3 million persons over 65 in the workforce; the Social Security Administration forecasts that by 2010 there will be about 5

Table 2–1. Population and Age Structure in the United States, 1950–2080 (Population in thousands. Includes armed forces overseas.)

Year	Total population	Age (years)									
		Under 5	*5–13*	*14–17*	*18–24*	*25–34*	*35–44*	*45–64*	*65 and over*	*85 and over*	*100 and over*
ESTIMATES											
1950	152,271	16,410	22,424	8,444	16,075	24,036	21,637	30,849	12,397	590	(NA)
1955	165,931	18,566	27,925	9,248	14,968	24,283	22,283	33,507	14,527	776	(NA)
1960	180,671	20,341	32,965	11,219	16,128	22,919	24,221	36,203	16,675	940	(NA)
1965	194,303	19,824	35,754	14,153	20,293	22,465	24,447	38,916	18,451	1,082	(NA)
1970	205,052	17,166	36,672	15,924	24,712	25,323	23,150	41,999	20,107	1,430	(NA)
1975	215,973	16,121	33,919	17,128	28,005	31,471	22,831	43,801	22,696	1,821	(NA)
1980	227,704	16,457	31,080	16,139	30,347	37,593	25,881	44,493	25,714	2,271	25
1982	232,057	17,372	30,431	14,963	30,367	39,481	28,144	44,574	26,824	2,445	32
PROJECTIONS											
Lowest series:											
1985	237,605	18,046	29,581	14,691	28,628	41,662	31,913	44,555	28,528	2,673	36
1990	245,753	17,515	31,638	12,848	25,547	43,147	37,570	46,136	31,353	3,202	50
1995	251,876	16,193	32,193	13,932	23,347	39,887	41,500	51,699	33,127	3,811	66
2000	256,098	14,942	30,364	14,587	24,157	35,596	42,972	59,859	33,621	4,444	84
2010	261,482	14,298	26,525	12,814	24,605	35,511	35,554	75,626	36,547	5,486	140
2030	257,443	12,136	23,686	11,369	20,454	30,434	34,679	66,600	58,085	6,490	233
2050	232,222	10,553	20,437	9,573	17,391	27,337	29,537	61,057	56,336	11,088	408
2080	191,118	8,479	16,360	7,725	14,107	21,919	23,863	49,630	49,035	10,085	551

Middle series:											
1985	238,631	18,453	29,654	14,731	28,739	41,788	32,004	44,652	28,608	2,696	37
1990	249,657	19,198	32,189	12,950	25,794	43,529	37,847	46,453	31,697	3,313	54
1995	259,559	18,615	34,436	14,082	23,702	40,520	41,997	52,320	33,887	4,073	77
2000	267,955	17,626	34,382	15,381	24,601	36,415	43,743	60,886	34,921	4,926	108
2010	283,238	17,974	31,888	14,983	27,655	36,978	36,772	77,794	39,196	6,551	221
2030	304,807	17,695	33,018	15,153	26,226	37,158	40,168	70,810	64,580	8,611	492
2050	309,488	17,665	35,583	14,600	25,682	38,383	38,844	74,319	67,412	16,034	1,029
2080	310,762	17,202	31,650	14,316	25,296	37,237	38,222	73,748	73,090	18,227	1,870
Highest series:											
1985	239,959	18,888	29,801	14,796	28,881	42,092	32,104	44,748	28,650	2,697	37
1990	254,122	20,615	32,935	13,120	26,137	44,329	38,229	46,767	31,989	3,379	57
1995	268,151	20,815	36,626	14,364	24,233	41,672	42,870	52,953	34,618	4,289	88
2000	281,542	20,530	38,128	16,306	25,326	37,850	45,128	62,025	36,246	5,387	136
2010	310,006	22,910	38,407	17,201	30,624	39,318	38,801	80,680	42,067	7,755	340
2030	369,775	26,562	46,999	20,567	34,190	45,739	46,278	76,854	72,587	11,417	1,016
2050	427,900	30,940	54,242	23,158	39,085	55,136	52,196	90,399	82,744	23,415	2,485
2080	531,178	37,439	65,466	28,236	47,911	66,393	63,744	112,094	109,896	32,456	5,932

Note: NA = Not Available.

Source: U.S. Bureau of Census, *Projections of the Population of the United States, by Age, Sex, and Race: 1983–2080*, Series P–25, No. 952 (Washington, D.C.: GPO), Table E, p. 7.

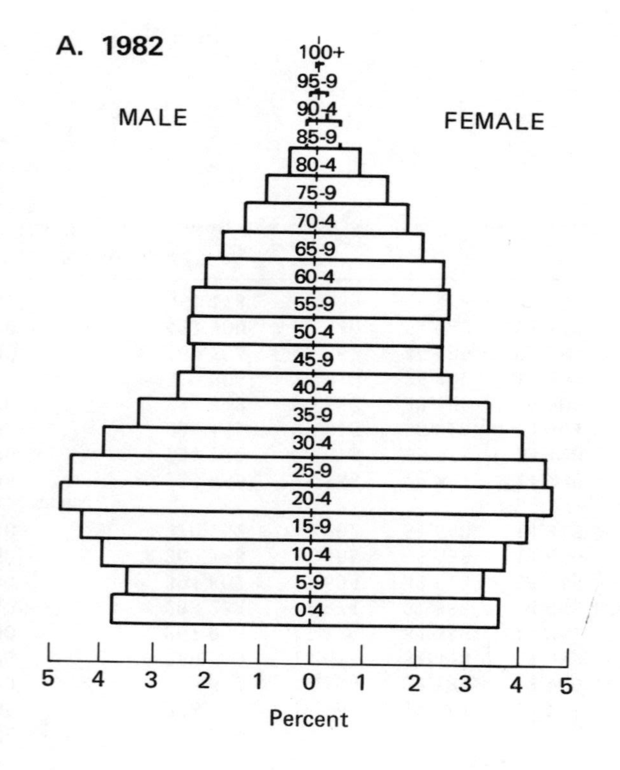
A. 1982
MALE
FEMALE
100+
95-9
90-4
85-9
80-4
75-9
70-4
65-9
60-4
55-9
50-4
45-9
40-4
35-9
30-4
25-9
20-4
15-9
10-4
5-9
0-4
5 4 3 2 1 0 1 2 3 4 5
Percent

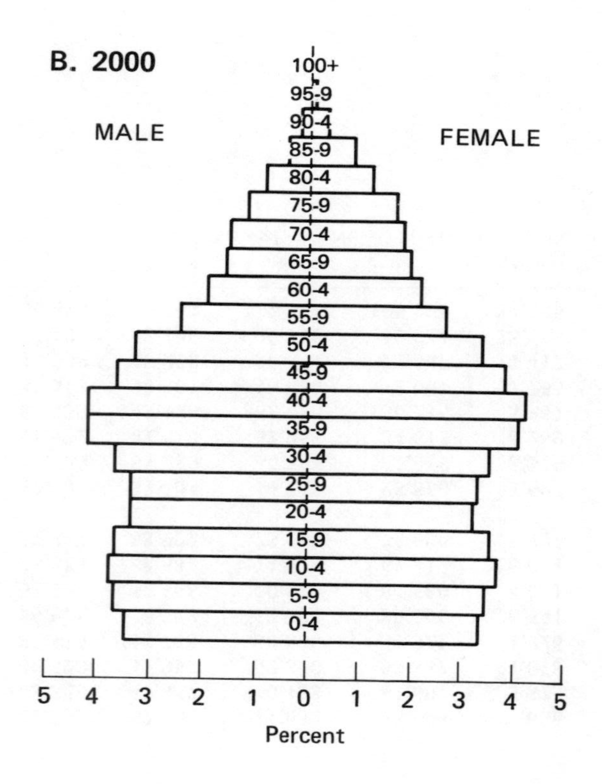
B. 2000
MALE
FEMALE
100+
95-9
90-4
85-9
80-4
75-9
70-4
65-9
60-4
55-9
50-4
45-9
40-4
35-9
30-4
25-9
20-4
15-9
10-4
5-9
0-4
5 4 3 2 1 0 1 2 3 4 5
Percent

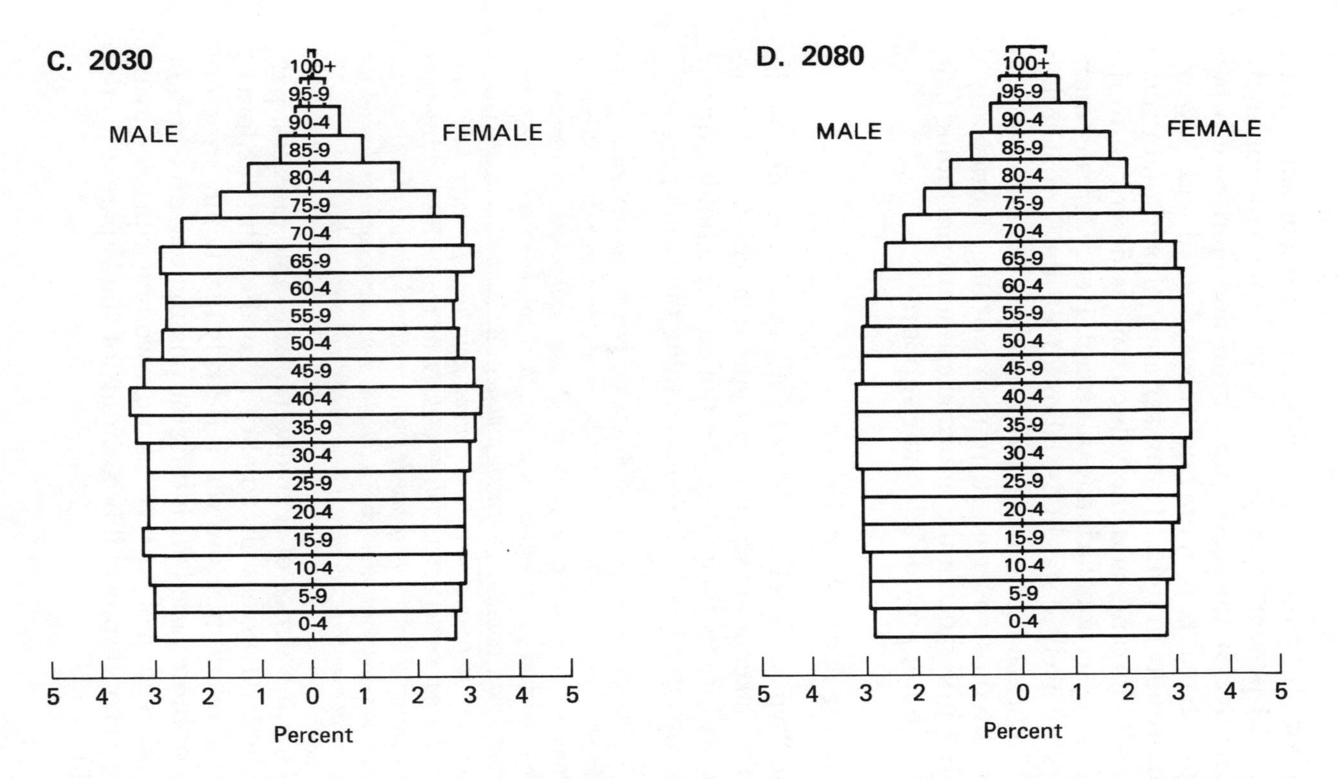

Figure 2-1 Changes in Age Distribution, 1980-2080. Source: Bureau of the Census, *Projections of the Population of the United States by Age, Sex, and Race: 1983 to 2080*, Series P-25, No. 952 (Washington, D.C.: GPO, 1984), p. 5.

million. For 65- to 69-year-olds, this represents an increase from 21 percent to 26 percent in their participation in the workforce. In the years ahead, more and more elderly persons will realize they are not "elderly" and do not have the opportunity to retire. The uncertainty of pension or social security income, and the value changes described in the following section, facilitate this trend. Hence, greater numbers are likely to stay in the labor force—the formal and/or the informal economy. There is likely to be ongoing pressure by those over 65 to remain working, particularly if public programs or the informal economy (including families) do not provide adequate support.

Changing Family Structure

The "typical family," with a husband wage earner, a wife homemaker, and two or more dependent children, now accounts for less than 10 percent of all households. Morton Darrow, in a study of trends shaping the family, argues:[3]

> *. . . though over 90 percent of Americans presently marry, by 2000 this may drop to 85 percent as many of the recent changes take hold. Stemming from the weakening of religious, social, and legal taboos, greater sexual freedom will promote continued growth of cohabitation, single-person households, unwed single-parent families, and homosexual couples. Over the next few years, despite the moral objections [of those with more traditional values], there will be widespread recognition of a family as consisting of two ore more people joined together by bonds of sharing and intimacy. To these two bonds is added the bond of commitment through the marriage contract, no matter how easy divorce is made.*

The prevalence of sexually transmissable diseases, particularly AIDS, has already begun to affect the freedom of sexual relations. However, it is likely that family structure will be more varied, including the possibility of even greater "feminization of poverty" as more women with low-paying jobs carry responsibility as heads of single-parent households.

Table 2-2. Labor Force and Labor Force Participation Rate by Age, 1980–2010 (in thousands)

	1980		*1990*		*2000*		*2010*	
Age	*Labor Force*	*Labor Force Partcpn. Rate*	*Labor Force*	*Labor Force Partcpn. Rate*	*Labor Force*	*Labor Force Partcpn. Rate*	*Labor Force*	*Labor Force Partcpn. Rate*
16–19	9549	57.7	8513	61.6	10035	67.1	10654	68.0
20–24	16192	78.1	15127	81.6	14731	84.7	16921	84.6
25–29	15114	80.6	17866	85.5	15605	87.9	17483	88.1
30–34	13719	79.9	17686	84.6	15569	87.3	14810	87.6
35–39	11171	80.1	16527	84.6	18082	87.5	15207	87.4
40–44	9335	80.2	15197	85.2	18949	88.2	16570	88.2
45–49	8498	77.3	11520	83.2	16698	86.4	17678	86.1
50–54	8465	73.0	8882	77.7	14280	81.9	17126	81.4
55–59	7269	64.4	6985	67.7	9161	70.3	12554	68.8
60–64	4488	46.3	5063	49.0	5537	54.3	8405	54.0
65–69	1811	21.1	2395	24.6	2248	25.9	2920	26.3
70+	1210	7.9	2030	10.3	2204	9.6	2130	9.0
Total > 16	106821	64.2	127796	68.3	143101	71.0	152460	69.7

Source: U.S. Social Security Administration, *Economic Projections for OASDI Cost Estimates, 1983*, Publication Number 11–11537, Actuarial Study No. 90, Washington, D.C.: SSA, 1984. Labor force estimates are from Alternative I, Table 12D, p. 70. Labor force participation rates are from Alternative I, Table 10E, page 56.

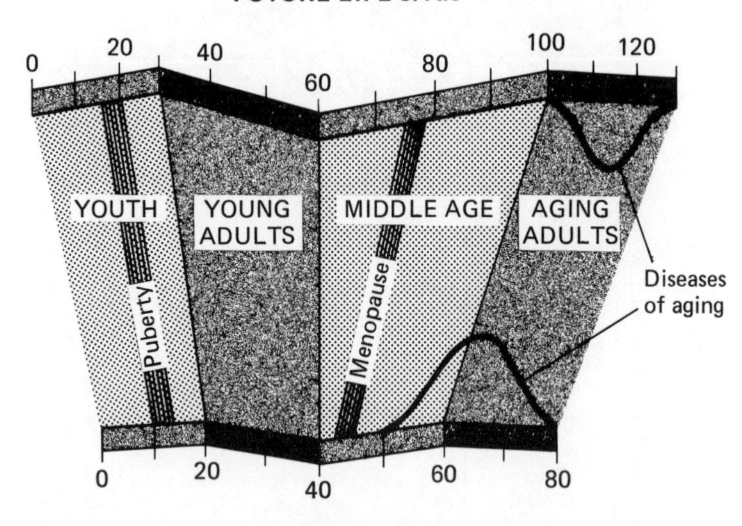

Extending maximum life span will stretch out the young-adult and middle-aged periods, probably with less extension of the period of decline. Longer youth period and later menopause will allow greater leeway in family planning. The diseases of old age will be delayed, and exposure to them will cover proportionately fewer years of the life span than they now do.

Figure 2-2 Comparison of Present and Future Life Spans. Source: Roy L. Walford, *Maximum Life Span* (New York: W.W. Norton and Co., Inc.), p. 190. Copyright © 1983 by Roy L. Walford.

The role of families and homes in education, including health and promotion of health, is likely to be aided by the variety of information devices that will become available over the next twenty-five years. Also, the ways in which families relate to the informal economy will be important determinants of their role in the formal workforce. Men are taking a more active role in parenting, and some of the caring functions which were taken from the extended family, put in the nuclear family, and then into the single-parent family may be reextended to neighborhoods and community and other local networks. (Forecasts for revitalization of the local informal economy are described in Chapter 3.)

Morbidity and Mortality

Developments around two important issues related to mortality and morbidity trends could affect the basic U.S. population forecast of 283 million by 2010. The first is life extension—the capacity to extend life beyond its natural limits. The second is the "compression of morbidity"—the capacity to ensure that most people live in a healthy condition until the natural limit of life. Life extension has received much popular attention, although there is a great deal of debate about its feasibility and, for some, its desirability as a goal of public policy. In Figure 2–2, Roy Walford[4] identifies how life span might be extended, including our sense of what constitutes youth and aging. Similar issues will be raised for work and promotion of health in the workplace more immediately in the consideration of the compression of morbidity.

Will the "compression of morbidity," assessed in more depth in Chapter 4 as a health trend, that has occurred in recent decades continue or even accelerate over the next twenty-five years? Briefly, the argument is that if personal behavior and certain allocations of resources were altered to emphasize prevention of premature death and disability, there would be significantly greater numbers of elders living relatively healthier lives up to a point much closer to their death. If this occurred, not only would there be an increase in elders relative to other age cohorts, as all current forecasts predict, but many more of these elders would be healthier. Hence, they would be more fit and able for work, and very possibly much more motivated to do so.

Immigration

Immigration patterns in this century have varied widely, as indicated by Figure 2–3. While the Census Bureau forecasts

Figure 2-3 Immigration Patterns, 1820-1970. Source: U.S. Bureau of the Census, *Statistical Abstract of the United States: 1984*, 104th Edition (Washington, D.C.: GPO, 1983), p. 89.

between 450,000 and 750,000 per year, many argue that the influx is more likely between 1 and 1½ million immigrants per year. At these higher figures, whites become a minority in the late part of the twenty-first century. Based on these data, and given the long-term decrease in fertility (taking into account the present "echo" of the baby boom, which is causing a short-term increase in the fertility rate), it is clear that the bulk of the projected population growth in the United States, at least in the near term, will arise from immigration, and most of that from Asian and Latin American countries. During the next ten years, Hispanics, as a group, will come close to becoming the largest minority, replacing blacks. Brown and Weiner note that "historically, immigrants have provided the backbone of change, innovation, and revitalization in the United States."[5]

It can be argued that the Census Bureau's forecasts are already too low and that pressures for immigration from Latin American countries are likely to increase because the percentage of young people in Latin American populations will dramatically increase. Without equally dramatic economic growth, there will be too few jobs in Third World economies to absorb the large numbers of younger people, which could increase immigration even more. The formulation of public policy affecting immigration laws is likely to have some impact on future immigration, but if recent history is a guide, illegal immigration into the Southwest is likely to persist, notwithstanding public policy. Also, conflicts and disruptions in other parts of the world might result in substantially greater demand for admission by immigrants, demands which this country has historically honored. Alternatively, the United States might choose and be able to effectively close its borders.

The Underclass

The future minorities in the United States and its labor force will be affected by immigration and differential fertility patterns. One of the major issues in the literature on work is the

Table 2–3. Females and Blacks Continue to Be Employed in Low-Paid Occupations

Occupation	*Percentage of Total Who Are Female*		*Percentage of Total Who Are Black*	
	1960	*1984*	*1960*	*1984*
Total labor force	33	44	11	10
White-collar	42	55	4	7
Professional and technical	36	48	4	7
Managers, officials, and proprietors	16	34	3	5
Sales	40	48	2	5
Clerical	68	80	5	10
Blue-collar	15	18	12	11
Craftworkers and foremen	3	9	5	7
Operatives	28	26	12	14
Nonfarm laborers	2	18	27	15
Service workers	65	61	27	18
Private household	98	96	50	30
Farm workers	18	16	16	8

Sources: Sar A. Levitan and Clifford M. Johnson, *Second Thoughts on Work* (Kalamazoo: The W.E. Upjohn Institute for Employment Research, 1982), p. 136, based on Employment and Earnings Report, January 1982, pp. 165–166 and 1981 Employment and Training Report of the President, pp. 149 and 151; and U.S. Department of Labor, *Employment and Earnings*, January 1985.

persistence of racial problems and inequities for blacks, particularly in two areas: a disproportionate and continuing low level of jobs, and unemployment. Sar Levitan provides Table 2–3 to show that between 1960 and 1981 there was slight change in the percentage of blacks in higher-paying jobs. Although some gains were made by women, blacks and women continue to be employed in low-paying occupations. "The distribution of jobs in the economy remains skewed to the detriment of blacks and women . . . . Blacks, though comprising only 11 percent of the labor force, hold 15 percent of all operative, 18 percent of laborer, and 20 percent of service jobs."[6] Brown and Weiner note that while official figures show a steady climb in earnings by black males as a percentage of white male earnings, these figures leave out the unemployed, masking a growing black male underclass.[7] Thus, lingering problems among blacks may worsen in the years ahead, as immigrants repeat the experience of recent decades: taking over the lowest-paying jobs and driving a corresponding increase in black unemployment.

Geographic Shifts

Much has been made of the so-called shift to the sun belt. Factors shaping this shift include better climate and lifestyle possibilities, lower wage rates, lower taxes, and other favorable economic conditions. These may be the central reasons for the documented shifts, but projections for the future based on these assessments are problematic, primarily because vital resources, especially water, are likely to decrease in quality and/or availability. Hence, the range in assumptions about the geography of work should be fairly wide, including, as some suggest, a renaissance of parts of the frost belt or rust bowl because of cōmpetitive wages, availability of water, and lower density than formerly desirable areas in the sun belt.[8]

Workforce Size

How many people will be working in 2010? The formal labor force projections from the Bureau of Labor Statistics (BLS) go through 1995 in published form[9] and to the year 2000 for total labor force in unpublished draft form.[10] Under the middle growth path forecast, the civilian labor force aged 16 and over increases from 110.3 million in 1981 to 137.8 million in 2000, as shown in Table 2–4.

The forecasts in Table 2–4 assume a steady pattern of U.S. economic growth, increasing 3.2 percent per year through 1990, and 2.5 percent per year after that, coupled with about 3.1 percent annual growth among the other nations of the world, a rebound in U.S. manufacturing productivity, and a

Table 2–4. Civilian Labor Force Participation Rate and Employment for Those Aged 16 and Over

	1981	*1985*	*1990*	*1995*	*2000*
Civilian labor force (in millions)	110.3	118.6	126.5	131.4	137.8
Participation rate	64.2%	65.9%	67.3%	67.8%	68.0%

Source: BLS/Norwood, Letter from BLS Commissioner Janet L. Norwood, September 20, 1984, pp. 2, 3, 36, 36.

slowing of the rate of growth of the service sector because of "maturation" (demand slackens with saturation). Loosening of tight government budgets allows more hiring; high-tech electronic fund transfer lowers employment in banking; and strong demand emerges for aerospace products and machinery in general.[11] Trends discussed in other chapters, particularly the health of the economy, the state of the informal economy, new technologies, and patterns of values, could significantly adjust these forecasts.

In tracking the basic federal government assumptions for labor force size, the Social Security Administration's forecasts go out further and also include a range of alternative forecasts. Table 2–2, based on assumptions most favorable to social security (higher labor force participation with slightly delayed retirement), indicates an increase in the labor force to 152.5 million by 2010 (the lowest estimate of the four developed in this Social Security Report was a labor force of 141.1 million by 2010). Thus, the current "official assumptions" see a labor force between 141 and 153 million in 2010.

Workforce by Sector

Given a workforce of a certain size, how will it be distributed in the future? In Table 2–5 the New York Stock Exchange has taken BLS data and separated the goods-producing sector from the service sector, showing that between 1962 and

Table 2–5. Distribution of Civilian Jobs, 1962–1995

Percent Distribution	*1962*	*1967*	*1972*	*1977*	*1982*	*1985*	*1990*	*1995*
Goods Producing Sector	57.3	52.8	50.2	47.9	45.7	45.0	44.3	42.8
Agricul., Forestry, Fisheries	7.8	5.3	4.3	3.6	3.5	2.8	2.4	2.0
Mining	.8	.7	.6	.7	.7	.7	.8	.9
Construction	5.6	5.1	5.5	5.3	5.4	5.9	6.2	6.2
Manufacturing	25.7	25.1	22.6	21.1	19.2	17.9	16.7	15.3
Non-Durables	11.2	10.3	9.5	8.7	7.7	6.8	5.7	4.8
Durables	14.5	14.8	13.1	12.4	11.5	11.1	11.0	10.5
Wholesale and Retail Trade	17.4	16.6	17.2	17.2	16.9	17.7	18.2	18.4
Total	57.3	52.8	50.2	47.9	45.7	45.0	44.3	42.8
Services Sector	42.7	47.2	49.8	52.1	54.3	55.0	55.7	57.2
Transportation	4.0	3.6	3.3	3.2	3.1	2.8	2.5	2.2
Communications	1.2	1.2	1.3	1.3	1.4	1.2	1.0	.8
Utilities	1.0	1.0	.9	.9	1.0	1.0	.9	.9
Finance, Insurance and Real Es.	4.7	4.5	5.0	5.2	5.6	5.9	6.0	6.2
Personal and Business Services	17.8	21.8	23.1	25.2	27.7	29.0	31.0	33.0
Civilian Government	14.0	15.1	16.2	16.4	15.5	15.1	14.3	14.1
Federal Government Enterprises	1.1	1.1	1.0	.9	.8	.7	.6	.6
State & Local Government (1)	10.2	11.3	12.8	13.3	12.7	12.3	11.7	11.5
Federal Government-Civilian	2.7	2.7	2.3	2.2	2.0	2.1	2.0	2.0
Total	42.7	47.2	49.7	52.2	54.3	55.0	55.7	57.2
TOTAL	100.0	100.0	99.9	100.1	100.0	100.0	100.0	100.0

NOTES

1. In the NYSE report, the projections for 1985 to 1995 for State & Local Government and the Federal Government were reversed. They are shown corrected, here.

Source: New York Stock Exchange, *U.S. International Competitiveness: Perception and Reality* (New York: NYSE, August 1984), pp. 32–44. Used with permission.

Table 2-6. Occupational Distribution of the Personal and Business Services Sector (millions of jobs)

	1982	*Percent Change*	*Absolute Change*	*1995*
Professional and technical	16.8	+36.3%	+6.0	22.8
managers, officials, proprietors	9.5	+33.4	+3.2	12.7
Sales workers	7.1	+32.0	2.3	9.4
Clerical workers	19.1	+31.8	+6.0	25.1
Craft workers	12.0	+28.8	+3.4	15.4
Operatives	12.9	+9.7	+1.2	14.1
Service workers	16.8	+35.4	+5.9	22.7
Laborers	6.0	+19.1	+1.1	7.1
Farmers and farm workers	2.8	−27.0	−0.8	2.0
Total	102.8	+27.9%	+28.7	131.5

Note: Details may not add due to rounding.

Source: New York Stock Exchange, *U.S. International Competitiveness: Perception and Reality* (New York: NYSE, August 1984), p. 46, Table 23. Used with permission.

1995 they reverse their positions (the goods-producing sector has 57.3 percent of the jobs in 1962, the service sector will have 57.2 percent in 1995). The NYSE definition of the service sector, unlike some others, excludes wholesale and retail trade on the argument that they are selling goods rather than the more intangible services. However, most of the change has already taken place in this area, so there is only slight adjustment between now and 1995 among the two. Again, Table 2-5 makes several assumptions that will be challenged in the following pages, but it does serve as a valuable starting point.

Job Classification and Types of Jobs in the Future

A critical question about the future of work concerns the nature or stature of the jobs that will exist. In Tables 2-6 and 2-7, the New York Stock Exchange had the University of Maryland break up the BLS data shown above by job type.

Table 2-7. Job Growth by Occupation, 1982 vs. 1995 (percentage of jobs)

	1982	*1985*
Professional and technical	27.6%	28.7%
managers, officials, proprietors	7.7	8.1
Sales workers	1.7	1.8
Clerical workers	17.3	17.7
Craft workers	5.8	5.8
Operatives	4.4	4.2
Service workers	33.0	31.1
Laborers	2.5	2.6
Farmers and farm workers	0.0	0.0
Total	100.0%	100.0%

Source: New York Stock Exchange, *U.S. International Competitiveness: Perception and Reality* (New York: NYSE, August 1984), p. 47, Table 22. Used with permission.

The NYSE results are hopeful. They write that "despite popular notions, more upper-echelon jobs will be added than lower-echelon jobs."[12] The contrasting "gods and clods" view of the future of jobs will be considered in Chapter 3. Table 2-6 provides the number and distribution of job growth between 1982 and 1995; Table 2-7 provides percentages for the same data, indicating relatively little change by 1995. Figure 2-4 portrays this data graphically, showing that there will be almost 6 million new jobs in each of the professional, technical, clerical, and service work fields, providing balanced growth across the job spectrum.[13] Figure 2-5 is the final aspect of the NYSE report, the percentage of professional and technical jobs in each sector. Expert computer systems and artificial intelligence will be used by and are likely to displace many of these workers, particularly in government and personal and business services.

The NYSE/BLS forecasts make several assumptions about society, the economy, and the nature of work and health. Several forces are addressed in Chapter 3 and 4 to affirm or alter the basic labor force assumptions identified in

Millions of Jobs	Occupation	% Growth
6.0	Professional and Technical	36.3
3.2	Managers, Officials and Proprietors	33.4
2.3	Sales Workers	32.0
6.0	Clerical Workers	31.8
3.5	Craft Workers	28.8
1.3	Operatives	9.7
5.9	Service Workers	35.4
1.2	Laborers	19.1
–0.8	Farmers and Farm Workers	–27.0

Figure 2-4 Job Growth by Occupation, 1982 to 1995. Source: New York Stock Exchange, *U.S. International Competitiveness: Perception and Reality* (New York: NYSE, August 1984), Chart 15, p. 47. Used with permission.

the forecasts discussed here. But before turning to other trends in the environment, it is relevant to review what is known about the size of the worksites for the U.S. labor force.

Worksite Size

What is the distribution of worksites by size for the U.S. workforce? There is no single accurate compilation of the data to answer this question fully. What is available are the data given in Table 2–8, which the Bureau of Labor Statistics compiled from state agency reports on unemployment compensation. There are some shortcomings in the data. For example, employers having a number of similar units within a given county are aggregated, including food stores and banks; thus some of the service, retail trade, and finance sector groups might show smaller units. But since the data in Table 2–8 cover about 70 percent of the workforce, focus-

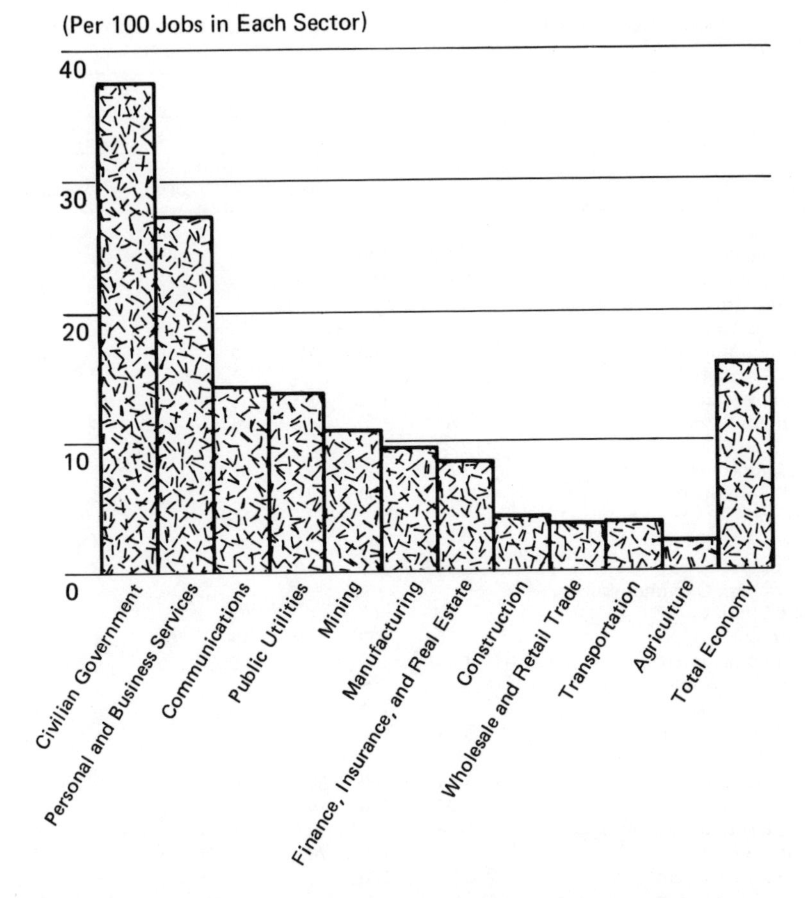

Figure 2-5 Professional and Technical Jobs by Sector, 1982. Source: New York Stock Exchange, *U.S. International Competitiveness: Perception and Reality* (New York: NYSE, August 1984), p. 47, Chart 14. Used with permission.

ing on private industry, it gives a sense of how worksite size is distributed. Thus, it indicates that in March 1983, 52 percent of all reporting units had 0 to 3 workers, even though they accounted for only 5 percent of the private industry workforce. At the same time, 40 percent worked in reporting units of 49 or less, 52 percent in units of 99 or less, and 67 percent in units of 249 or less. In the retail trade sector, 52

Table 2–8. Distribution of the United States Workforce in Private Industry by Size of Worksite, March 1983

			Size of					
	Total		*0–3*		*4–9*		*10–19*	
Industry	*Number (1,000s)*	*Pct. (%)*	*Number (1,000s)*	*Pct. (%)*	*Number (1,000s)*	*Pct. (%)*	*Number (1,000s)*	*Pct. (%)*
All industries								
Reporting Units	4,763	100	2,461	52	1,209	25	520	11
March Employment	71,631	100	3,508	5	7,057	10	6,980	10
Agriculture, Forestry, & Fisheries								
Reporting Units	111	100	61	56	28	26	12	11
March Employment	976	100	77	8	166	17	156	16
Mining								
Reporting Units	44	100	20	46	9	21	6	14
March Employment	944	100	25	3	56	6	83	9
Construction								
Reporting Units	485	100	297	61	109	22	44	9
March Employment	3,492	100	324	9	634	18	588	17
Manufacturing								
Reporting Units	342	100	95	28	77	23	55	16
March Employment	18,102	100	140	1	473	3	762	4
Transportation, Communications, & Public Utilities								
Reporting Units	188	100	85	45	46	24	25	13
March Employment	4,558	100	120	3	270	6	337	7
Wholesale Trade								
Reporting Units	468	100	223	48	124	26	64	14
March Employment	5,187	100	320	6	735	14	862	17
Retail Trade								
Reporting Units	1,147	100	501	44	346	30	151	13
March Employment	15,053	100	790	5	2,038	14	2,027	13
Finance, Insurance & Real Estate								
Reporting Units	403	100	245	61	84	21	33	8
March Employment	5,294	100	349	7	481	9	443	8
Services								
Reporting Units	1,544	100	909	59	381	25	128	8
March Employment	17,878	100	1,333	7	2,173	12	1,702	10

*Percentages may not add to 100 due to rounding.

Source: U.S. Bureau of Labor Statistics. Based on reports from state unemployment insurance agencies.

Unit

20–49		*50–99*		*100–249*		*250–499*		*500–999*		*1,000 and Over*	
Number (1,000s)	*Pct. (%)*	*Number (1,000s)*	*Pct. (%)*	*Number (1,000s)*	*Pct. (%)*	*Number (1,000s)*	*Pct. (%)*	*Number (1,000s)*	*Pct. (%)*	*Number (1,000s)*	*Pct. (%)*
348	7	123	3	69	1	20	1	8.3	1	5.2	1
10,549	15	8,459	12	10,451	15	6,858	10	5,656	8	12,115	17
6	6	1	2	.8	1	.2	0	.07	0	.01	0
187	19	125	13	129	13	76	8	45	5	16	2
4	11	1	4	1	2	.3	1	.2	0	.07	0
147	16	125	13	159	17	113	12	109	12	126	13
25	5	6	1	2	1	.5	0	.1	0	.09	0
742	21	438	13	356	10	151	4	75	2	185	5
53	16	27	8	21	6	7	2	3.4	1	2.0	1
1,664	9	1,851	10	3,203	18	2,675	15	2,290	13	5,046	28
19	10	6	4	4	2	1	1	.5	0	.5	0
586	13	464	10	607	13	425	9	357	8	1,392	31
41	9	11	2	4	1	.9	0	.3	0	.08	0
1,226	24	755	15	680	13	294	6	181	3	133	3
100	9	31	3	13	1	3	1	1.1	0	.6	0
3,026	20	3,104	14	1,946	13	1,023	7	772	5	1,327	9
24	6	9	2	5	1	1	0	.6	0	.4	0
741	14	629	12	768	15	512	10	405	8	967	18
73	5	28	2	17	1	4	0	2.0	0	1.5	0
2,208	12	1,954	11	2,589	14	1,584	9	1,417	8	2,917	16

percent of employees work in sites of 49 or less. In this sector all units of a particular chain (McDonald's restaurants, for example) in the same county aggregate their employees, so the percentage of workers in small sites in this sector is likely to be higher than these data reflect.

Notes

1. *Business Week*, "A Work Revolution in U.S. Industry," May 16, 1983, pp. 10–110.
2. Fred Best, "Recycling People: Work Sharing Through Flexible Life Scheduling," *The Futurist* (February 1978, pp. 5–16; see also the 1984 and 1985 issues. Published by The World Future Society, 4916 St. Elmo Ave. Bethesda, Md. 20814.
3. Family Service America, *The State of Families 1984–85*, New York, 1984, p. 7.
4. Roy L. Walford, *Maximum Life Span* (New York: Avon, 1984), p. 191.
5. Arnold Brown and Edith Weiner, *Supermanaging* (New York: McGraw-Hill, 1984), p. 29.
6. Sar A. Levitan and Clifford M. Johnson, *Second Thoughts on Work* (Kalamazoo, MI: W.E. Upjohn Institute for Employment Research, 1982), p. 140.
7. Brown and Weiner, p. 33.
8. *Ibid.*
9. Bureau of Labor Statistics, *Employment Projections for 1995,* Bulletin 2197 (Washington D.C.: GPO, March 1984).
10. ______, Janet L. Norwood, BLS Commissioner, Correspondence and forecasts to the year 2000 (forecasts unpublished) (Washington, D.C.: Bureau of Labor Statistics).
11. Bureau of Labor Statistics, Bulletin 2197, pp. 22–23.
12. New York Stock Exchange, *U.S. International Competitiveness: Perception and Reality* (New York: NYSE, Office of Economic Research, August 1984), p. 46.
13. *Ibid,* p. 47.

Chapter 3

KEY TRENDS SHAPING THE FUTURE OF WORK AND THE WORKPLACE

Major trends are shaping the future of work and the workplace in terms of the economy, demographics, values, technology, and the worker and the workplace. Let us review each of these areas and look at the implications.

The Economy

The Rise of the Service and Information Economy

A much discussed trend is the shift from a predominantly manufacturing economy and workforce to one which is largely services-oriented. Related arguments have been made that this is really a shift to an information economy. Figure 3-1 charts that dramatic rise. As John Naisbitt writes, we have gone from a society of predominantly farmers to one of factory workers to one of predominantly clerks. Thus today's typical worker moves, handles, and transfers—and even occasionally analyzes and interprets—information as opposed to minerals or auto parts.

One aspect of this information economy is the "disembodiment" of information processing. Information technologies in many cases are doing jobs that traditionally have been done manually. Another cluster of issues that bears on forecasts about the service and information economy arise from pending policy debates about "reindustrialization." If the United States embarked on what some call reindus-

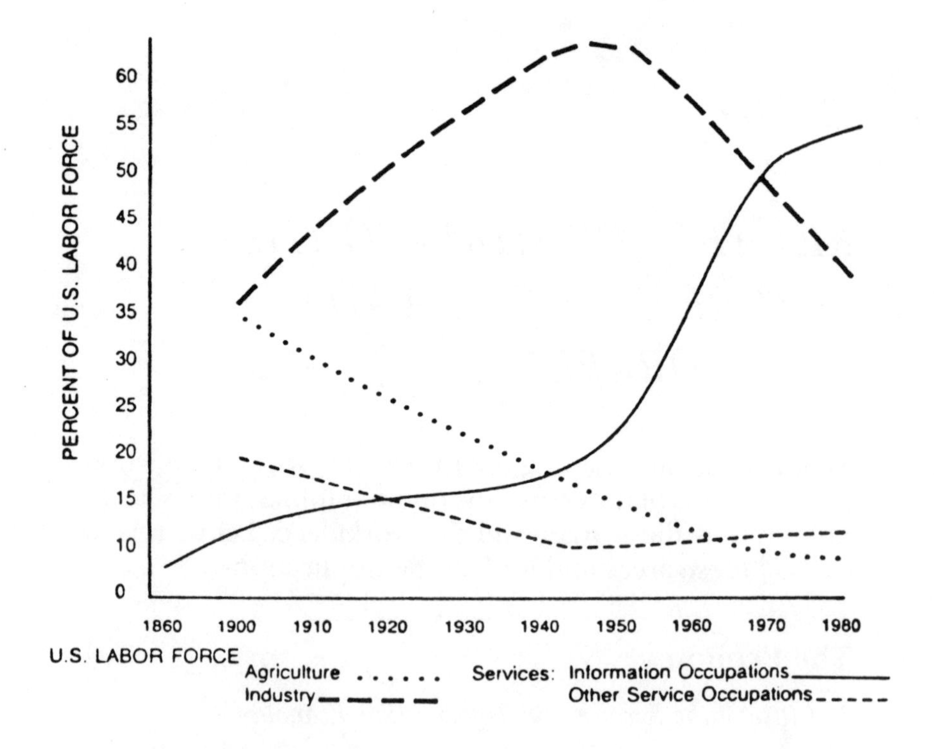

Figure 3-1 The Rise of Information Workers. Source: J.F. Coates, "The Changing Nature of Work," in *The World of Work*, Howard F. Didsbury, ed. (Bethesda, Md.: The World Future Society, 1983), p. 26. Used with permission.

trialization,[1,2] the number of jobs forecast for the manufacturing sector might increase relative to those in the service sector. On the other hand, the extent to which the next generation of industrial technologies would depend on—and stimulate—the information economy is difficult to establish. A new industrial infrastructure would likely produce a different balance between the number of workers involved in production and the numbers needed for distribution of products (see the discussion of the role of new technologies on job creation later in this chapter). One aspect of the changing economy that has not received as much attention is the growing information value or content of the products we buy. Paul Hawken, in *The Next Economy,* argues that products are becoming more "information rich" to meet the growing consumer demand for products that intelligently fit their particular needs.[3]

Whatever the definition, and however imprecise the term, there is no doubt that this trend is likely to grow along with the capacities of our information technologies. As Eli Ginzberg points out, "The current revolution in computer technology is causing an equally momentous social change: the expansion of information gathering and information processing as computers extend the reach of the human brain."[4]

Computers will increasingly allow machines to augment and enhance the functioning human brain for a wide variety of applications. Thus, in terms of work, there are potential problems with the growth of the service/information economy. Service jobs are traditionally low paying, and for many jobs computers can facilitate relatively low levels of skill and therefore keep pay low. Computers also can replace many service personnel—bank tellers, for example. Forecasts for loss and displacement of jobs to computers have existed for more than two decades. In 1962, for example, Donald Michael, in *Cybernation: The Silent Conquest,* forecast a loss of jobs in some areas, massive displacement in others, and the need to prepare for changes in jobs during the coming decades.[5] The first wave of computers increased total employment because they created jobs and, in many cases, made little change in the way work was done. Now some forecasters argue that jobs will be significantly altered, if not lost, when computers and robots become more sophisticated, approaching "expert systems" or "artificial intelligence."

For our purposes, three issues are important. First, as a society are we becoming more dependent on services? Some services are a natural outgrowth of the number and nature of other jobs and the changing nature of the workforce. For example, as the workforce continues to include

Table 3–1. The Twenty Fastest Growing Occupations, 1982–1995

Occupation	*Percent growth in employment*
Computer service technicians	96.8
Legal assistants	94.3
Computer systems analysts	85.3
Computer programmers	76.9
Computer operators	75.8
Office machine repairers	71.7
Physical therapy assistants	67.8
Electrical engineers	65.3
Civil engineering technicians	63.9
Peripheral EDP equipment operators	63.5
Insurance clerks, medical	62.2
Electrical and electronic technicians	60.7
Occupational therapists	59.8
Surveyor helpers	58.6
Credit clerks, banking and insurance	54.1
Physical therapists	53.6
Employment interviewers	52.5
Mechanical engineers	52.1
Mechanical engineering technicians	51.6
Compression and injection mold machine operators, plastics	50.3

Note: Includes only detailed occupations with 1982 employment of 25,000 or more. Data for 1995 are based on moderate-trend projections.

Source: U.S. Bureau of Labor Statistics, *Employment Projections for 1995*, U.S. Department of Labor, March 1984, Bulletin 2197, p. 44.

more women, more two-career families, longer hours for some, and flextime and shorter hours for others, these changes continue to create growing demands for food service workers, child or day-care services, counseling for drug and alcohol abuse, stress-management seminars, tax preparers, and the like.

Second, it is unlikely that the more routine service jobs will gain either the status of more traditional manufacturing or industrial jobs or the commensurate compensation and benefits associated with them. The third issue is that if the economy declines or if large portions of the workforce are

Table 3-2. The Twenty Most Rapidly Declining Occupations, 1982–1995

Occupation	*Percent decline in employment*
Railroad conductors	−32.0
Shoemaking machine operatives	−30.2
Aircraft structure assemblers	−21.0
Central telephone office operators	−20.0
Taxi drivers	−18.9
Postal clerks	−17.9
Private household workers	−16.9
Farm laborers	−15.9
College and university faculty	−15.0
Roustabouts	−14.4
Postmasters and mail superintendents	−13.8
Rotary drill operator helpers	−11.6
Graduate assistants	−11.2
Data entry operators	−10.6
Railroad brake operators	−9.8
Fallers and buckers	−8.7
Stenographers	−7.4
Farm owners and tenants	−7.3
Typesetters and compositors	−7.3
Butchers and meatcutters	−6.3

Note: Includes only detailed occupations with 1982 employment of 25,000 or more. Data for 1995 are based on moderate-trend projections.

Source: U.S. Bureau of Labor Statistics, *Employment Projections for 1995*, U.S. Department of Labor, March 1984, Bulletin 2197, p. 444.

left with no work or only low-paying jobs, the market for services such as fast-food preparation and child-rearing services could increasingly move out of the formal economy. They will be picked up by families or through noncash alternatives.

Employment and Unemployment

What is the level of employment and unemployment likely to be over the next twenty-five years? As already noted, the Social Security Administration expects relatively little un-

Table 3–3. The Forty Occupations with the Largest Job Growth, 1982–1995

Occupation	*Change in total employment (in thousands)*	*Percent of total job growth*	*Percent change*
Building custodians	779	3.0	27.5•
Cashiers	744	2.9	47.4•
Secretaries	719	2.8	29.5
General clerks, office	696	2.7	29.6
Salesclerks	685	2.7	23.5
Nurses, registered	642	2.5	48.9•
Waiters and waitresses	562	2.2	33.8
Teachers, kindergarten and elementary	511	2.0	37.4
Truckdrivers	425	1.7	26.5•
Nursing aides and orderlies	423	1.7	34.8•
Sales representatives, technical	386	1.5	29.3
Accountants and auditors	344	1.3	40.2
Automotive mechanics	324	1.3	38.3•
Supervisors of blue-collar workers	319	1.2	26.6•
Kitchen helpers	305	1.2	35.9•
Guards and doorkeepers	300	1.2	47.3
Food preparation and service workers, fast food restaurants	297	1.2	36.7•
Managers, store	292	1.1	30.1
Carpenters	247	1.0	28.6•
Electrical and electronic technicians	222	.9	60.7•
Licensed practical nurses	220	.9	37.1•
Computer systems analysts	217	.8	85.3
Electrical engineers	209	.8	65.3
Computer programmers	205	.8	76.9
Maintenance repairers, general utility	193	.8	27.8•
Helpers, trades	190	.7	31.2•
Receptionists	189	.7	48.8•
Electronics	173	.7	31.8•
Physicians	163	.7	34.0
Clerical supervisors	162	.6	34.6
Computer operators	160	.6	75.8
Sales representatives, nontechnical	160	.6	27.4
Lawyers	159	.6	34.3
Stock clerks, stockroom and warehouse	156	.6	18.8
Typists	155	.6	15.7
Delivery and route workers	153	.6	19.2
Bookkeepers, hand	152	.6	15.9

Cooks, restaurants	149	.6	42.3•
Bank tellers	142	.6	30.0
Cooks, short order, speciality and fast food	141	.6	32.2•

Note: Includes only detailed occupations with 1982 employment of 25,000 or more. Data for 1995 are based on moderate-trend projections.

Bulleted (•) occupations are among those currently highly susceptible to injury or disease.

Source: U.S. Bureau of Labor Statistics, *Employment Projections for 1995*, U.S. Department of Labor, March 1984, Bulletin 2197, p. 43.

employment, with between 141 and 152 million persons in the workforce in 2010. Tables 2–6 and 2–7 in Chapter 2 identify the types of jobs in general, forecasting that 28.7 percent of the workforce of 1995 will be in the professional and technical category, while 17.7 percent will be clerical and 31.1 percent will be service workers. Table 3–1 shows the twenty fastest growing occupations from 1982–1995, and Table 3–2 shows the twenty most rapidly declining occupations during that same period, according to BLS. However, in terms of the overall pattern of jobs, Table 3–3 shows the forty occupations with largest job growth, led by janitors, cashiers, secretaries, and salesclerks and office clerks.

The BLS data have been criticized for extrapolating current assumptions and thus neglecting changes in technology that make the forecasts inaccurate. For example, the projected increase in the number of salesclerks does not take into account the automated tellers, which, after reducing the need for more bank tellers, may replace a percentage of the sales force projected for settings such as department stores. Likewise, the 96.8 percent growth projected for computer service technicians (see Table 3–1)ignores the increasingly sophisticated development of hardware and software, which can reduce the need for human maintenance. The microcomputer industry is also developing new office equipment that will merge voice and data technology in workstations, which may decrease the number of secretaries

Table 3-4. Office and Factory Jobs Affected by Computers and Robots

	Number of Employees
In Factories	
Assemblers	1,289,000
Checkers, examiners, inspectors, testors	746,000
Production painters	185,000
Welders and flame-cutters	713,000
Packagers	626,000
Machiner operatives	2,385,000
Other skilled workers	1,043,000
Total	6,987,000
In Offices	
Managers	9,000,000
Other professionals	14,000,000
Secretaries and support workers	5,000,000
Clerks	10,000,000
Total	38,000,000

Source: *The Impacts of Automation on the Workforce and Workplace*, Carnegie-Mellon University; Booz, Allen & Hamilton, Inc. Cited in Sar Levitan and Clifford M. Johnson, *Second Thoughts on Work* (Kalamazoo: The W.E. Upjohn Institute for Employment Research, 1982), p. 108. Used with permission.

and clerks needed. Other technologies, such as self-diagnosis kits, may reduce the need for workers who are now needed to perform older, more complex diagnostic tests in central labs. As Richard Bolles puts it, "hand, eye and head jobs will continue to decline in number." It is difficult to project when and to what extent technologies will have an impact on job markets. However, it is clearly dangerous to rely solely on extrapolative job forecasts without looking at related technologies that are developing.

Futurist and author Marvin Cetron notes that "some economists are predicting that during the next 20 years office automation will eliminate between 30 and 50 million additional jobs from a white-collar work force that numbers about 90 million."[6] He predicts a "structural unemployment" rate of 8.5 percent in the 1990's, up from about 4.5 percent in 1984. (Structural unemployment is the expected

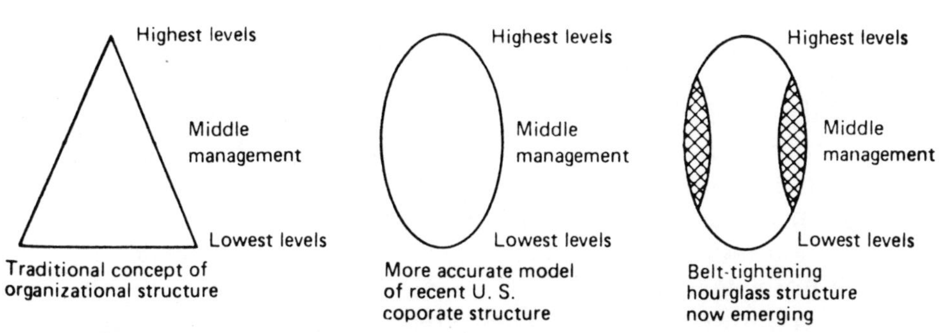

Figure 3-2 The Emerging Organizational Structure. Source: Arnold Brown and Edith Weiner, *Supermanaging* (New York: McGraw-Hill, p. 99). Used with permission.

number of persons out of work because of continuing changes in the economy—the number that, even with full employment, would still be out of work.) Sar Levitan likewise forecasts, as shown in Table 3–4, that 7 million manufacturing and 38 million white-collar jobs will be "affected" by computers and robots. He cites the Carnegie-Mellon study as noting that the 38 million jobs will be "affected" by computers but not eliminated.

Others think that the second wave of computers will have an effect similar to the first wave—namely, an overall creation of jobs. John Naisbitt argues that unemployment will not be a problem. "From about 1986, we'll have full employment and tremendous labor shortages for the rest of the century. As we are now approaching the end of the maturing of the baby boom, the economy is accelerating and creating new jobs. In less than three years from now, there will be a negative net gain—more people will be leaving than joining the workforce, which will continue for the rest of the century."[7] Levitan says there will be an ample supply of labor and of jobs, and while the requirements for jobs will not change dramatically, new technology will generate lower-skilled and lower-paying jobs in far greater numbers than higher-skilled jobs.[8]

Arnold Brown and Edith Weiner see more change in the nature of work. In particular, they see the thinning of the ranks of middle management because of economic, demographic, and technological developments. They use the diagram in Figure 3–2 to illustrate the coming shifts of organizational form from one of an oval with a large middle to something closer to an hourglass. If this organizational thinning takes place, it will be across the leading edge of the baby boom generation. These changes will take place at different rates in different industries; there has been little change in textiles and other traditional manufactures, for example, while other businesses are bringing the top and bottom closer together. For example, Equitable, a major insurance company, has recently undergone a significant reduction in the levels of gradation among its employees, tied to a greater decentralization of decision-making among its operations, and a shift to incentive, performance-based pay. Some companies are experimenting with no more than three levels of employees below the president of a company. The federal government is using similar programs to shift incentives to workers and, in at least one setting, is experimenting with a reduction of the current eighteen grades of pay down to four.

There are divergent yet co-existing trends here: One makes personnel in organizations more equal and gives them more incentives for productive performance; another gives a job-surplus-and-worker deficit; and yet another largely eliminates mid-level jobs. In this latter possibility, there are fewer middle-level jobs and some high-paying jobs, but the latter are relatively few in relation to less well paying jobs, particularly in the growing service sector. Thus some analysts have forecast a "gods and clods" image of the future of the job market, one with a relatively small percentage of high-paying jobs and a vast majority of low-paying jobs. Sam Cole and Ian Miles have reviewed the impact of the microelectronics field and argue that not only will these mid-level jobs decline but the purchasing power of the middle level will decline proportionately and have persistent negative effects on the economy.[9] This has not been adequately studied.

This decline may complement the movement of manufacturing jobs abroad that is reshaping the formal economy. If this is true, it becomes important to ask whether these trends presage a possible economic decline, or whether the informal economy combined with the rise of the service and information economy can sustain a growing workforce. The decline is likely to be pronounced for those blue-collar workers who still have jobs. Arthur Shostak, a Drexel University sociology professor and futurist, argues that blue-collar workers will have substantially reduced disposable income and purchasing power by the year 2000.[10] For society, this may mean either the emergence of a large number of impoverished unemployed along with a smaller number of employed blue-collar workers or a reconceptualization of work.[11]

Globalization

Today we live in an increasingly transnational global network of economic transactions. The global economy is not limited to just a few industries or technologies; there is competition from abroad for the bastions of U.S. economic activity such as automobiles, transport, plastics, machine tooling, steel, and so on. Much of this competition arises because of the supply of cheap labor abroad. For example, in the U.S. auto and steel industries, labor costs are more than $20 dollars an hour, whereas in Mexico, workers earn only about a third of the U.S. minimum wage of $3.35 an hour, and in Singapore and Hong Kong much manufacturing labor costs only $1.50 an hour.[12] This imbalance drives manufacturing jobs abroad, making the 22 percent of the American workforce in manufacturing jobs extremely vulnerable to global competition. A much smaller percentage of workers in the service sector, on the other hand, face competition from abroad. Yet even the service area faces competition. For example, American Airlines now processes data in Barbados.

While high-technology jobs are sometimes touted as America's answer to the loss of manufacturing jobs, there is also real and fierce competition in many of the new "high-tech" fields such as computers, robotics, and microelectronics. This competition will be strongly affected by such factors as the overvalued American dollar. The globalization of currency, banking, and other financial transactions includes massive foreign investment in the United States (and some foreign investors are very volatile) and U.S. loans overseas (some of which are very uncertain).

By 2010 it is likely that many of the factories which moved overseas will return to the United States since robots will become cheaper to use than foreign labor and since the costs associated with political instability and transportation for Third World production will rise. Some industries are already experiencing these impacts. While "it was once 25 percent cheaper to make a basic white shirt abroad than in the United States, the gap has now narrowed to 10 percent."[13]

Possible Economic Decline

The major forces shaping the economy are discussed throughout this chapter. There are questions about the nature of the information revolution, the growth of the service sector, and other aspects of the "next economy." Generally, they are optimistic in nature. Yet the future is uncertain, and it is therefore relevant to ask what might be some of the major negative trends that could affect the economy over the next twenty-five years. Leaving aside the prospects for major or minor nuclear war and natural calamities of high impact, among the leading possibilities for a significant economic downturn are a global financial collapse, leading to a global depression;[14] a rapid outflow of foreign capital from the United States; or, some argue, immutable forces in the economy which could cause a significant downturn. These latter "long wave" or "Kondratieff wave" theorists pose a plausible, if undesirable, challenge for speculation on the future of work and health.

Most economic forecasting presumes a pattern of stable and substantial economic growth, although with possible fluctuations in the economic cycle that can be "corrected" by government intervention. Moreover, economic forecasts are also based on the assumption that history repeats itself (the good parts at least) and that only the recent past economic experience is relevant to predicting the future. However, there is a substantial body of theory, opinion, and some evidence that the economy goes through longer-term cycles of growth and decline. These cycles consist of upwaves and downwaves of growth which occur with some regularity. The "long-wave theory" is most associated with the work of the Russian economist Nicolai Kondratieff, although his work has been refined by Schumpeter, Mensch, and more recently by Jay Forrester of MIT. The theory, boldly stated by Robert Beckman in his book *The Downwave,* is "based on the idea that there is a strong interaction between political and social developments, wars and the long-term price cycle, all of which come to a peak at 50–60 year intervals."[15]

Hank Koehn has examined data for banks, businesses, and other sectors of the economy and has concluded that it is uncertain if we are presently in an economic upswing, a plateau period, or heading into a serious decline.[16] This muddled picture is the result of the mixture of economic trends underway today, including unsteady interest rates, moderate to high levels of unemployment, increasing levels of investment and savings, the large federal deficit, and so forth. One charting of these cycles appears in Figure 3–3. Obviously, if major shifts occur in the status of the economy of the magnitude reflected in Figure 3–3, most existing economic forecasts and, in turn, estimates of the size of the workforce will be relatively useless for part of the 1980's and 1990's. By 2010 the economy would be about 5 to 10 years into its upswing.

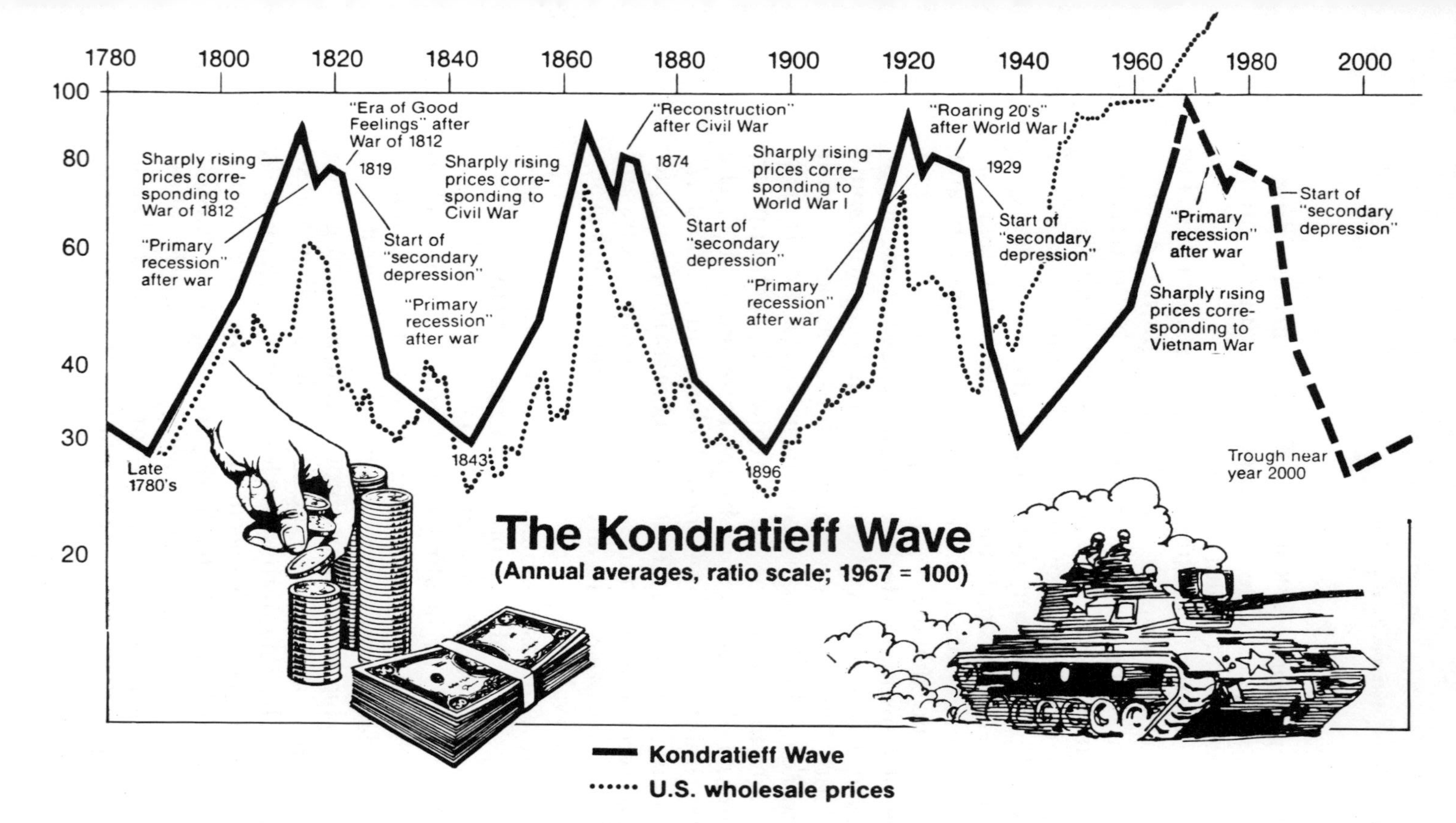

Figure 3-3 Long Wave Cycles in the U.S. Economy. Source: Valerie Andrews, "Boom or Bust: Economic Cycle or Human Sacrifice?" *The Tarrytown Newsletter,* January 1984, p. 8. Used with permission.

Decentralization

As pointed out in Naisbitt's *Megatrends,* one of the most significant and unmistakable long-term trends in the United States is decentralization, reversing the powerful centralization dynamic that has characterized the United States since shortly after World War I. Decentralized decision-making, and even discretion over resource allocations, has begun to occur in the business community. According to Rosabeth Moss Kanter, in *Change Masters,* decentralization is gradually becoming recognized as "a broad condition for innovation in a company."[17] Kanter cites Hewlett-Packard and 3M, although very large companies, as pioneers in decentralized decision-making, including some resource allocation decisions. Yet there will continue to be strong movements in corporations against noncentral control, at least in part arising from concerns over the cost effectiveness of decentralization. Alternatively, some companies may be slow to change, and many will be forced to maintain high levels of control as they shrink their workforces because of hard times or rapid change in their business environment. The key questions relate most directly to the possible impacts of these developments on the nature and quality of work, the scale and quality of environment in the workplace, and on job formation.

Decentralization and regional and local self-sufficiency movements outside the workplace introduce other dimensions. There are those who argue in "Ecotopian" terms (to borrow Ernest Callenbach's term) that decentralization and greater self-sufficiency not only will improve the quality of work and the quality of health in the workplace, but that *our society as a whole* will be healthier. Recent conceptual arguments have been provided by Jane Jacobs in her book, *Cities and the Wealth of Nations: Principles of Economic Life*, stating that cities and city regions are the real creators of wealth and they can choose to be more self-reliant and therefore import and export less.[18] And David Morris argues that "self-reliant" cities make much better use of their resources.[19] The Cornucopia Project at Rodale Press is using similar terms to analyze the potential for greater bioregional self-reliance at the state level. This trend toward decentralization is likely to result in the shifting of activity to the informal local economy.

Increased Entrepreneurial Activity

Alongside and related to the restructuring of businesses, there is a trend toward increased entrepreneurial activity, particularly as the service sector of the economy grows over manufacturing. Entrepreneurs may take over more and more of the innovation functions which are the key to market growth. The United States is a world leader in providing customer satisfaction; fueled by the influx of capital from abroad the marketing of new services, including entrepreneurial consulting through groups such as the American Corps of Entrepreneurs, could achieve explosive growth in the new economy. Douglas Greene, founder of the American Corps of Entrepreneurs and owner-developer of his own publishing company, argues that the new entrepreneurs will focus on needs; they will be more sensitive to potential new markets than to expansions of existing products. They will seek to become what Bill Gore, the founder and owner of the company which produces Gore-tex fabric for outdoorwear, calls "heroes," people who meet others' needs effectively and imaginatively. These entrepreneurs may work in a number of different businesses simultaneously. Because the key to opening new markets is entrepreneurial risk-taking, the dynamic growth in the economy will center on the activities of these entrepreneurs, primarily in smaller businesses.

While the success and glorification of the entrepreneur is a cause for elation among some, it raises concern among others, who argue that perhaps 10 percent of the population may have a bright future, while those below the median continually suffer from changes in the economy and social fabric. In a world where successful entrepreneurs become cultural heroes, there may also be "psychic inflation" as Faustian images replace the coherent social bonds of a more traditional workplace. Others argue that there is not conclusive evidence that this trend toward entrepreneurship is taking place.

Disintermediation

Related to decentralization, disintermediation is a trend that is particularly important for jobs in the years ahead. This term, used by Paul Hawken in *The Next Economy,* refers to the elimination of the middle-man or of intermediate services provided in certain settings. These settings are increasingly familiar. They include telephone services; gasoline pumping; and wholesale purchase of toys, clothing, and even groceries, as in the case of wholesale and/or "warehouse" grocery shopping. At one level, disintermediation is generally inconsistent with the growth of the service sector, since the inference is that certain service jobs, like telephone operators and clerks, are made unnecessary. It is difficult to predict how widespread this phenomenon is.

Disintermediation is also conceptually related to the observation by some that a growing number of consumers believe themselves to be sufficiently competent to make a wider range of choices and decisions about goods and services than historically has been the case. As noted by Naisbitt in *Megatrends,* Ogilvy and Mather, the world's third largest advertising agency, reached this conclusion in an extensive study of the attitudes and behaviors of TV viewers toward commercials.

Of course, competencies differ—the exercise of competence to pump your own gas is different from the competence it takes to treat a cancer. The hard question is where to place limits on forecasting this trend. Brown and Weiner argue that "individuals are looking for a way to exert control over their own lives—to resist external control by large organizations and institutions."[20] Expressions of this desire to exert control may be what Ogilvy and Mather picked up

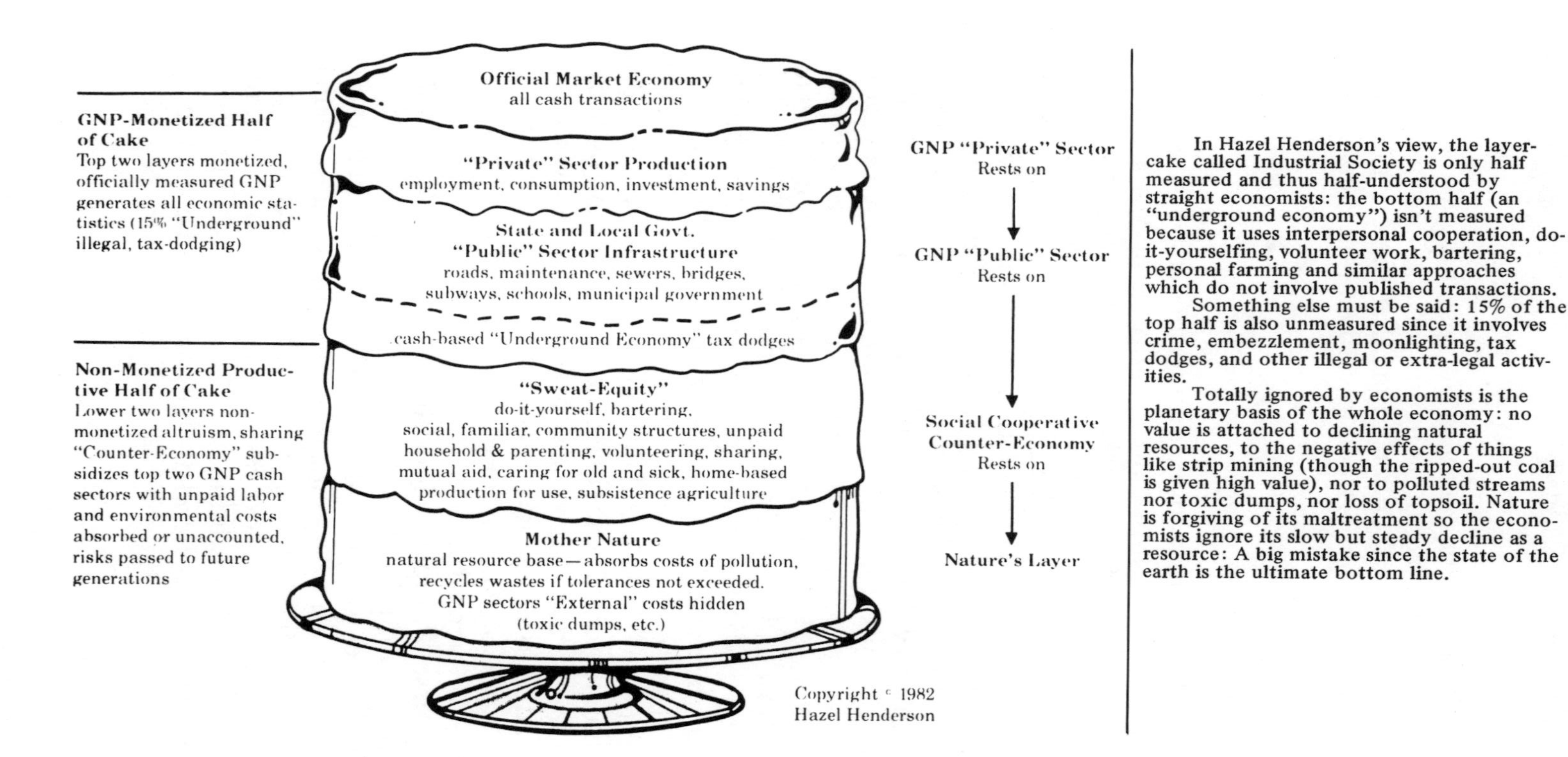

Figure 3-4 The Layercake Theory of Economic Interaction. Source: Valerie Andrews, "Hazel Henderson: Will the Real Economy Please Stand Up?" *Tarrytown Letter*, January 1984, p. 6. Used with permission.

in their study. In the health trends described later in this chapter, they are related to the forces giving rise to self-care.

Disintermediation is related to but different from the "thinning of the middle" argument discussed earlier. The thinning of the middle focuses on manufacturing and service delivery organizations and states that more streamlined organizations will emerge. Disintermediation argues a compressing of the points of contact between consumers and producers of goods and services.

The Informal Economy

Disintermediation and decentralization are among the forces increasingly leading human exchange out of the formal economy. Virtually all measures of economic growth reflect only the formal or measured economy. In Figure 3–4, Hazel Henderson argues that the monetized economy is composed of the formal, counted economy *and* the uncounted underground cash economy. Both parts of the cash economy rest on the informal economy of individuals, families, and communities sharing and bartering their goods and services. The above three rest on the health of the natural/ecological system. Figure 3–5 illustrates the cash portion of these components. In policy discussions the cash portion of the informal economy is increasingly recognized, and, in most policy dicussions, it involves the middle two layers of Hazel Henderson's diagram. However, estimates of the size of this cash portion of the informal economy vary widely, from $118 billion to more than $500 billion a year. Although the figure used most often for the informal economy is about 10 to 15 percent of the GNP, is it likely to grow in the future?

The Environmental Scanning Committee of United Way of America, in developing four scenarios for the United States in the year 2000, predicted that the informal economy would range between 15 percent and 50 percent of the size of the formal GNP by the year 2000, depending on eco-

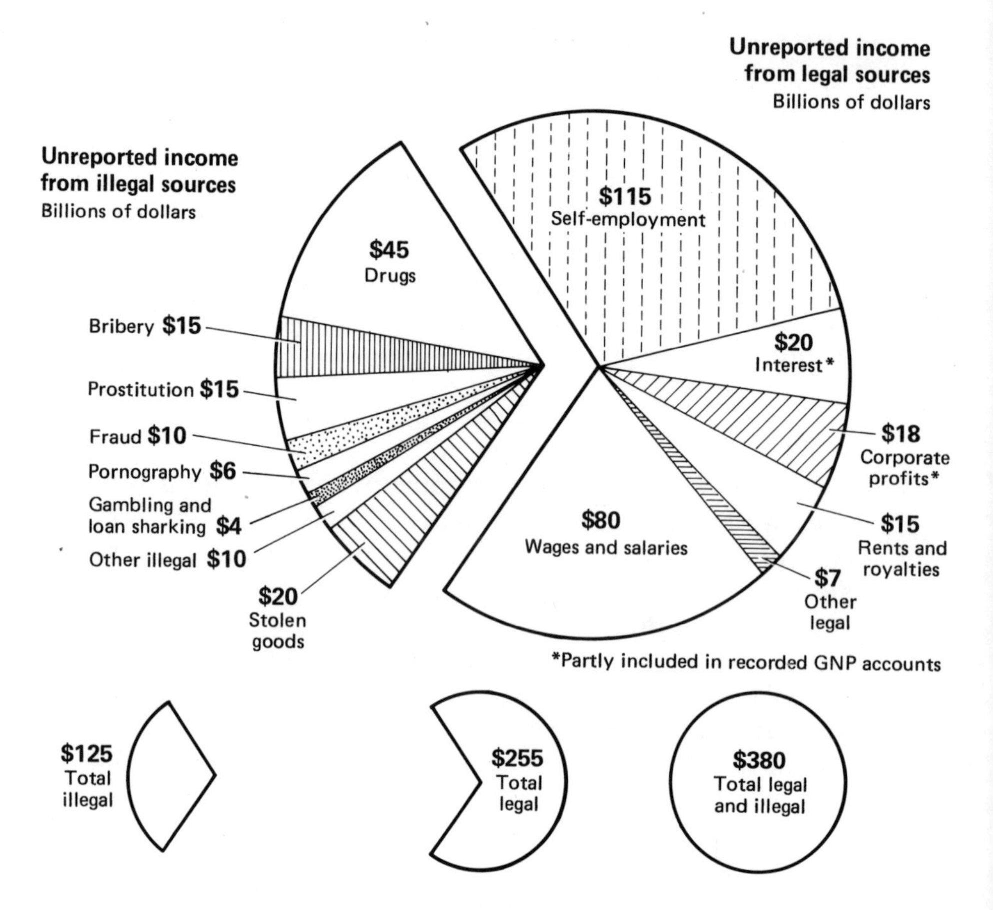

Figure 3-5 Sources of the Cash Portion of the Informal Economy. Source: Reprinted from the April 5, 1982, issue of *Business Week* by special permission. Copyright © by McGraw Hill, Inc.

nomic conditions and value shifts.[21] Whatever its actual size, however, even the most conservative estimates make clear its impact on economic life. Work done off the books has become a way of life for many Americans, and a common activity for even more.

Existence of this "economy" has profound implications for understanding our economy at large, its potential growth, and the nature of work in the future. If, for example, the rate of growth in the informal economy is significantly greater than in the formal economy, then the true growth of economic activity is much larger than generally presumed. Some argue that because of its size and relationship to the state of our society's economic health, society must find ways to value, compensate, and reward informal economic activity. Some have explained that the reason for so little social unrest during the 1981–82 recession, when unemployment reached the highest levels since the Great Depression, was precisely because the informal economy absorbed many of the unemployed.

The informal economy helps reveal the different social functions work performs. Work, beyond producing goods and services for society and distributing wealth, promotes the development of individuals and provides them with a healthy self-image.[22] The question that arises is whether the informal economy will absorb more of these functions while the formal economy becomes more narrowly (and efficiently) focused on producing goods and services. According to futurist James Robertson, this can lead to a merging of work and leisure that will see increasing numbers of people filling local needs outside the "cash" economy.[23]

New Definitions of Work: New Ways to Distribute Income

This subject is less a trend than an issue for public policy analysis. Nevertheless, if other trends identified in this book have some of the impacts suggested, a new definition of work and ways to distribute income other than jobs will emerge. For example, as Wassily Leontief points out in a recent article in *Scientific American,* whenever workers are displaced by machines—which is one plausible interpretation of current trends—then purchasing power is diminished, with all the attendant and rippling effects.[24] Cole and Miles, as noted earlier, argue that the microelectronics field will lead to diminished purchasing power overall.

A more unfortunate effect is the expansion of the underclass. To some observers, the new technologies will not only displace workers but leave only two classes of jobs: elite jobs, filled only by those who know how to program and manage the technologies, and service and maintenance jobs, which will be as poorly paid and little valued as they are now. Of course, welfare and welfare-type programs, including income maintenance, may result from society's beneficence. But unless destigmatized, the vision of the underclass forecast by James Robertson, Arnold Brown and Edith Weiner, and Sar Levitan is a likely one.

Values

Changing Values: Implications for the Economy and for Work

Extensive research indicates that throughout Western industrial countries major value shifts are taking place. These shifts are not always uniform for all U.S. workers, and in some cases they can be contradictory (for example, the rise of conservatism and, simultaneously, "expressive" values). The value shift most important for our discussion of work is what Daniel Yankelovich calls "expressive values." They are described in a major report by Yankelovich and an international team of experts for the Aspen Institute entitled *Work and Human Values: An International Report on Jobs in the 1980s and 1990s*.[25]

The core aspects of these expressive values are (1) defining success on the basis of inner growth rather than on external signs of wealth, (2) desiring to live in harmony with nature, (3) seeking autonomy and questioning authority, (4) greater freedom in seeking pleasure, and (5) yearning for stronger and closer bonds of community with others—from friends and neighbors to humankind.[26] Yankelovich notes that a variety of observers have discussed this value shift as part of a shift in societies from agricultural civilization to industrial civilizations and then to affluent societies undergoing major "transformations." Table 3–5 compares the terminology of these authors and their similar approaches.

These changing values are important both for what con-

Table 3–5. Changing Values

Author	*Values in Agrarian Societies*	*Values in Industrial Societies*	*Values in Post-affluent/ Transformational Societies*
Sorokin		Sensate	Ideational
Fromm		Having	Being
Inglehart		Materialist	Postmaterialist
Toffler	First Wave	Second Wave	Third Wave
Mitchell	Need driven	Outer-directed	Inner-directed
Yankelovich et al.	Sustenance	Material Success	Expressivism

Source: Adapted from Yankelovich, et al., *Work and Human Values: An International Report on Jobs in the 1980s and 1990s* (New York: Aspen Institute for Humanistic Studies, 1983), page 52.

sumers will seek from the marketplace and for the impact such a shift would have on work. Figure 3–6 identifies the movement of core values in Sweden from 1875 to 1965 as that nation moved from a predominantly agrarian society to an industrial one and, more recently, toward a third type. No similar chart exists for the United States. However, based on the research reviewed by Yankelovich and his colleagues, it appears there is a similar, although somewhat slower trend: 17 percent of U.S. workers exhibited expressive values in the early 1980's as opposed to 23 percent of Swedish workers.[27]

Work and Value Change

Work has always been judged to have a significant value, a value that is characterized both by extrinsic rewards (compensation, benefits, recognition, status, and so forth) and intrinsic rewards (such as self-satisfaction, personal achievement, opportunities for self-enhancement and self-improvement, a resulting happy family life, and others). To the extent this is true, attitudes and motivation toward work are bound to be a function of the value placed on it by the individual and by society. Similarly, productivity is associated with attitude, motivation, and desire for material success.

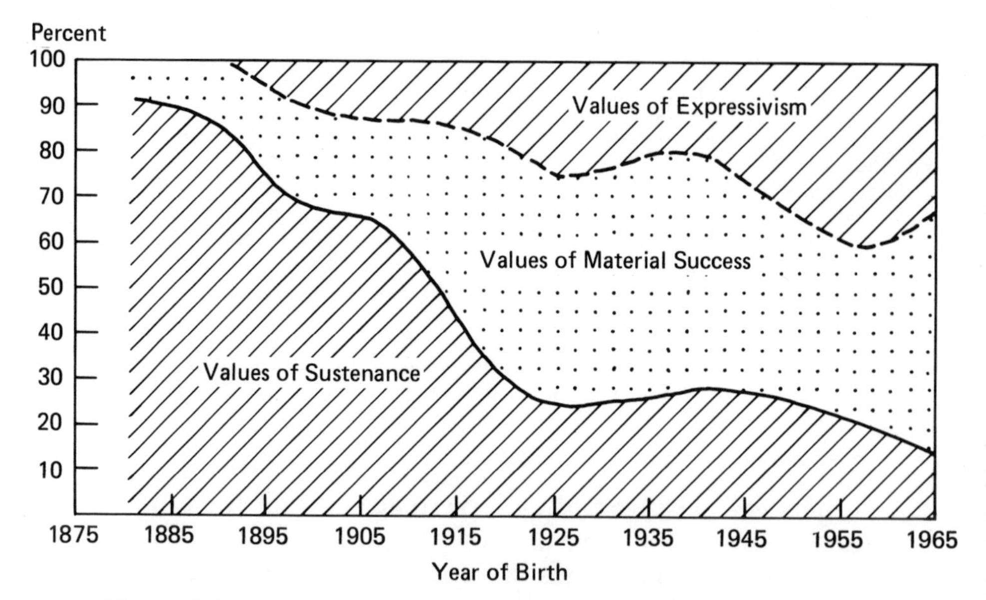

Figure 3-6 Change of Core Values in Sweden, 1875-1965. Source: Daniel Yankelovich, et al., *Work and Human Values: An International Report on Jobs in the 1980s and 1990s* (New York: Aspen Institute for Humanistic Studies, 1983), page 54. Used with permission.

How have workers' attitudes and motivations changed, and what is the relationship between those changes and productivity? Furthermore, given the changes in work and the workplace which are likely to occur, what are worker attitudes and motivations likely to be in the future? The following are the most important findings Yankelovich offers on the issue of values and attitudes:[28]

1. The recently documented declines in productivity in the United States are associated with larger changes in attitudes, characterized as "expressivism."
2. The report defines "expressivism" as including values like "creativity, autonomy, rejection of authority, placing self-expression ahead of status, pleasure-seeking, the hunger for new experiences, the quest for

community, participation in decision-making, the desire for adventure, closeness to nature, cultivation of self, and inner growth." And, as the report points out, "One of the most clear-cut findings of our research is the documentation of how extensively these new values have been transferred to the workplace."

3. One of the ways this "transfer" has taken place is that workers increasingly seek to express these values in their work as well as in other settings. As the report goes on to say, " . . . .the initial effect of new values in the 1960's and 1970's (in the context of existing corporate organization and philosophies) was to cause jobholders to demand more from the workplace in terms of rights and rewards while giving less in terms of effort, quality and commitment." Simple productivity measures often ignore, and may conflict with, expressive values. Will the workplace move away from productivity measures in response to a value shift?
4. Frequently, the response of management to these new values has been to bemoan losses in productivity and rebuff the demands for more participation. The report argues, however, that these new values are so entrenched that it is management that must adapt; the workplaces of the future must offer opportunities that meet these new expressive values. And this must be done at a time when economic pressure, including the heralded plans for "revitalization," usually cause management to streamline by cutting "labor costs." Nevertheless, the report states unequivocally, "The real challenge is to develop strategies that are not built around excluding people and reducing jobs, but that depend more on the full utilization of human resources."

Another important value identified by the Aspen report is a growing "respect" for quality. It is frequently observed that lower productivity may be associated with worker sloppiness, poor work habits, low job satisfaction, and so on.

The report and other sources show that job satisfaction is low and has been declining. It is also true that some of this dissatisfaction is due to the failure of the workplace to provide opportunities for new "expressive" values. As the report makes clear, however, to conclude that worker respect for quality and integrity of product or service itself has diminished would not be fair. On the contrary, productivity appears to suffer most when management evinces little respect for quality, and where little, if any, attention is given to worker participation in product improvement and worker pride in the product or the service.

James O'Toole makes a similar argument for the expressive values being those of the baby-boom generation, particularly the values of change, flexibility, choice, options, variety, and diversity. Current modes of corporate organization and philosophy of work do not tap the positive values of these workers; instead they encourage counterproductive traits of narcissism, irresponsibility, and selfishness. The changes in the organizational cultures in corporations that fit with the need for "responsibility, rights, quality, ownership, and participation" will be those which are most successful in the years ahead.[29]

The baby-boom generation will continue to be a major factor in working America well beyond 2010 (when the oldest of the baby boomers are reaching their early 60's). It is reasonable to expect their expressive values to have great influence as they move through the workforce. But will the expressive values of the baby-boom cohort be dominant? If, as Alvin Toffler argues, these value shifts are part of a larger change in civilization, then they are likely to emerge as prime shapers of change through society. If, as James O'-Toole argues, the value shifts are related to the life experiences of the baby boomers growing up in the 1950's and 1960's, then the 1970's and 1980's may be inculcating an as yet unfactored set of values into the younger members entering the workforce in the 1990's and the decade before 2010.

Worthiness of Products

The movement toward the "worthiness of products" as a criterion which will affect both production and consumption decisions is a controversial subject. Proponents argue that the market will become more consciously aware of the social effects of products, which will influence not only the enthusiasm and productivity of workers but also the buying patterns of consumers.

Arguments of this sort have been present for some time. There is some experience with workers doing long-range product planning based on a standard of product worthiness. For example, the workers at Lucas Aerospace in Great Britain developed a definition of "socially useful" products and generated a series of new product ideas which the company could produce, given its equipment and employees' skills.[30] Similar ideas have been promoted by various worker-planning and appropriate technology movements in the United States.

In his review of this point, John Naisbitt argued that companies that produce products of questionable worth will have more difficulty competing for the best talent. Similarly, futurist and management expert Harlan Cleveland argues that this is a call for ethical behavior that needs to be reflected in corporate management:[31]

> *Big corporations now usually have a vice president for keeping the corporation out of trouble with nosy outsiders, or even with their own stockholders and employees, who raise questions about what the company ought to produce, who it ought to employ, and how it ought to invest its money.*
>
> *Should "my" company, or any American company, make and market nerve gas, even if the government does want to buy some? . . . . Shouldn't "my" company have more women, and blacks, and American Indians in its employ? . . . Should "my" company, or any American company, pass the "social costs" of its profit-seeking—overcrowding, the paving of green space, radioactive risk, dirt, noise, toxic waste, acid rain, or whatever—to the general public?*
>
> *. . . These "public responsibility" issues can make or break companies,*

products, and executive reputations. If you don't beleive that, take a Nestle executive to lunch and ask him about marketing baby formula in the Third World.

. . . The visibly responsible leaders increasingly have to build into their organizations, not as a public relations frill but as an essential ingredient in "bottom line" budgeting, staff members competent to develop strategy on such issues as these.

On the investment side of the market, the "social investment" funds, such as the Calvert Social Investment Fund, are among the most successful of the money market funds. These firms use explicit criteria for determining the nature of the product and the safety and environmental quality of the production processes, and only organizations fulfilling the guidelines are eligible for funding. Likewise, some bankers argue that banks which have seldom raised these questions are beginning to look more closely at their business loans to companies involved in activities such as pornography.

As this changing set of values and greater concern for the quality of community and environment emerges, personal choices will lead to demands for better products. While this trend is less visible than others discussed here, it parallels the rise of more active consumer monitoring of health care. Knowing how "socially useful" a product or service is, and communicating that to consumers, will be made far easier by community computer networks using various "expert system" software programs.

Competing consumer groups with slightly different values are likely to emerge, encouraging consumers to clarify their own values. Consulting groups, such as the VALS program at SRI International and New Value Enterprises, already exist to translate the product implications of the new values into new product ideas for producers. Physician Russell Jaffe, working with New Value Enterprises, argues that companies in all sectors, but particularly the health sector, need to understand the emerging shift among consumers from a focus on quantity and price to a focus on value and quality.

Given this set of forces and the value shift we described, individuals are likely to seek more sophisticated standards for judging how products and their production processes meet their needs and enhance or retard the type of community conditions (and global conditions) consumers prefer. When markets shift, only a small percentage of the population need be active in order to have the marketplace read its signals. Thus, 5 to 20 percent of the population could provide an influential minority to shape the market.

Technology

Jobs from High Technology

It is patent that almost any type of technological development has implications for work and the workplace. This was true of the steam engine, the internal combustion engine, gunpowder, and the computer, and it will be true of the "new" technologies. The critical questions, then, are *what* technologies and what impact on jobs will the introduction of these particular technologies have on: (1) job formation; what type, directly or indirectly created; (2) job replacement; and (3) values about the nature of work and the relationship among work, income, social status, and the overall economy.

What Are the "New" Technologies?

There are many lists of such technologies. Hank Koehn, in "The Once and Future Economy," includes one of the most comprehensive: "microelectronics, computers, biotechnology, factory automation, lasers, space manufacturing, robotics, new energy sources, advanced materials utilization, surface science, holography, bionics, fiber optics, video, artificial intelligence, advanced information/communication technology."[32] Koehn also includes a list of the types of services in the service economy. They include

Table 3–6. Arthur C. Clarke's Candidates for Major Scientific and Technological Breakthroughs, 1800–2100

Date	*Transportation*	*Communication Information*	*Materials Manufacture*	*Biology Chemistry*	*Physics*
1800			Steam Engines	Inorganic Chemistry	Atomic Theory
	Locomotive				
		Camera			
		Babbage Calculator		Urea synthesized	
1850		Telegraph	Machine Tools		Spectroscope
					Conservation of Energy
			Electricity	Organic Chemistry	
		Telephone			
		Phonograph			Electromagnetism
	Automobile	Office Machines			Evolution
1900			Diesel Engine		
	Aeroplane		Petrol Engine	Dyes	X-Rays
					Electron
		Vacuum Tube	Mass Production	Genetics	Radio-Activity
				Vitamins	
1910			Nitrogen Fixation	Plastics	
					Isotopes
		Radio			
1920					Quantum Theory
				Chromosomes	
				Genes	
					Relativity
1930					Atomic Structure
				Language of Bees	
				Hormones	Indeterminacy
		TV			Wave Mechanics
					Neutron

1940	Jet				
	Rocket	Radar			
	Helicopter				
		Tape Recorders	Magnesium from	Synthetics	Uranium Fission
		Electronic Computers	Sea	Antibiotics	Accelerators
1950		Cybernetics	Atomic Energy	Silicones	Radio Astronomy
		Transistor	Automation		
		Maser			
	G.E.M.		Fusion Bomb	Tranquilisers	I.G.Y. Parity
	Satellite	Laser			Overthrown
1960	Spaceship			Genetic Code	Nuclear Structure
		Comsats			
1970	Moon Landing	Pocket Calculators		Organ Transplants	Pulsars
	S.S.T.				Strange Particles
1980	Space Probes	Video Recorders	Solar Energy	Gene Splicing	
	Shuttle		Ocean Thermal Power		
1990	Space Labs	Pocket Educators		Animal Languages	Gravity Waves
		Libraries	Electric Storage		
2000	Planetary Landings	Universal Radiophone	Sea Mining	Cloning	Super-Heavy Elements
			Non-Cryonic Super-	Consciousness	
2010	Earth Probes		conductors	Expansion	Magnetic Monopoles
		Tele-sensory Devices	Fusion Power	Suspended Animation	
2020	Hypersonic Transport	Artificial Intelligence			
				Intelligent Animals	Nuclear Catalysts
2030	Lunar Settlements	Detection of Extra-	Weather Control		
		Solar Intelligence		Brain Transplants	
2040	Deep Sea Probes				
		Memory Recording			
2050	Space Colonies		Space Mining	Artificial Life	
					Space, Time Control
2060		Mechanical Educator			

Table 3–6 *(continued)*

Date	*Transportation*	*Communication Information*	*Materials Manufacture*	*Biology Chemistry*	*Physics*
	'Space Drive'		Transmutation		
2070					
		Artifact Coding	Replicator		
2080	Planetary Colonies				
2090	Space Elevator	Machine Intelligence Exceeds Man's	Planetary Engineering		
2100					
Beyond 2100	Interstellar Flight First Contact	World Mind	Climate Control Cosmic Engineering	Immortality	Black Hole Experiments

Source: From *Profiles of the Future* by Arthur C. Clarke, p. 253. Copyright © Holt, Reinhart and Winston, New York, 1984 and reprinted by permission of the publisher.

"communication, transportation, food preparation, vacations, entertainment, information, health care, and consulting." He adds that "the entire financial, legal, investment and insurance communities are all service industries.[33]

The Role of Computers, Artificial Intelligence, and Robotics

Some jobs may be altogether displaced by a new technology. The use of robots, for example, represents a clear case of job displacement—they are designed to do just that. Although estimates of displacement vary, one possible scenario, offered by Vary T. Coates of J.F. Coates, Inc., a futures policy group, suggests, "it is certainly possible that within two to three decades we could see a million or more robots filling three million of today's jobs."[34] She adds that the job displacement might be as high as 40 to 50 million persons. (As noted, the Social Security Administration sees a labor force of 141 to 152 million in 2010—without considering this displacement.)

Let's turn back, then, to the hard question, What will be the various impacts of these new technologies and services? This report assumes that many of these new technologies will be inherently job displacing. For example, fifth-generation computers will greatly enhance the capacity of our machines to function like our brains. Also, our capability to develop knowledge bases—information plus the rules for manipulating that information—will be far more sophisticated. Thus, by 2010 developments in expert systems, the fifth generation of computers, and information compression and storage could displace many knowledge-dependent jobs. This includes much of what middle management does and what physicians and other professionals do, such as information procesing and paper or electronic manipulation of assets or resources (banking and investment, for example). Sam Cole and Ian Miles argue that microelectronics is likely to have an overall job-displacing effect.[35]

The Impact of Still Newer Technologies

It is hard to anticipate precisely which technologies discovered and developed over the next twenty-five years will affect work. Yet there is a clear sense that information technology, communication technology, and biotechnology are among the leading technologies that will make life and work in the United States different by the early part of the twenty-first century. To these three, Arnold Brown and Edith Weiner add materials technology, a field that is likely to give us polymers that can substitute for strategic minerals; superconductivity in devices that function at room temperature and provide for no energy loss while moving electricity; and new methods for joining metals and alloys that will allow more efficient use of existing resources and creation of wholly new products.[36]

In *The Technology Edge,* Gerard K. O'Neill suggests six categories of emergent technological development: neuroengineering, robotics, genetic manipulation, magnetics in transportation, private aviation, and space travel.[37] To remind us of breakthroughs in the past and provide his candidates for future breakthroughs, Arthur C. Clarke provides Table 3-6. Included among those available by 2010 are pocket educators and libraries, electric storage, noncryonic superconductors, cloning (as with Brown and Weiner), and consciousness expansion. Clarke sees full artificial intelligence available by 2020. (The advances described in the work and health trend sections as expert systems are close approximations to the functioning of our intelligence and are likely to have the impacts described even though they are not full analogues to the brain.)

The Worker and the Workplace

There are a host of trends affecting work and the workplace directly. Arnold Brown and Edith Weiner use Figure 3-7 to

Workplace Changes

Driving Forces	Flexible hours	Work at home	Support for networks	Stress management	Restructuring of jobs	New recruiting and promotional considerations	New compensation schemes	Professional contacts
Two-wage-earner households	X	X				X	X	
Broadened social and self-awareness	X		X	X	X	X		X
Explosive growth of service economy	X	X	X	X	X		X	X
Economic, demographic, and technological challenges to middle-management positions				X	X	X	X	
Time of rapid change and an uncertain future			X	X	X	X	X	X

Figure 3-7 Responses to Change in the Workplace. Source: Arnold Brown and Edith Weiner, *Supermanaging* (New York: McGraw-Hill, 1984), p. 101. Used with permission.

identify some of the driving forces and consequent workplace changes. We have just discussed the changes in the economy, technology, and values. This section deals with some of the changes taking place in the nature of work and workplaces, including work schedules, organizational culture, pensions and social security, workplace safety and health, viewing employees as "human capital," and the future of worker organizations.

Work Schedules

Work schedules are dramatically changing, not only in terms of when the work is done, but also regarding the length of the workweek, its traditional full-time nature, the work schedule's consistency over the lifetime, and even where the hours are put in. The length of the workweek has dropped over the last forty years, and many argue that it will decrease further still. For example, Marvin Cetron argues that "by 1990 the average worker will put in 32 hours a week and 25 hours a week by 2000."[38] The Bureau of Labor Statistics concurs but points out that this includes full-and part-time workers—it forecasts for nonfarm workers an average workweek of 33.1 hours by 1995, down from 35.1 in 1982 (based on both a decrease in full-time workers and an increase in part-time workers). Sar Levitan points out that the evidence is not definitive, and in his estimation the bulk of full-time workers have remained on a forty-hour workweek for decades; the overall decline has occurred because of the increase in part-time workers. By 2010, what percentage of full-time workers will be working less than, say, 34 hours a week? If large numbers of jobs are lost, some workers may be more ready to share, while others will not be.

A related matter is the frequency with which a worker changes careers or makes a major job shift. It is now estimated that most workers will perform five or six different jobs over the course of their careers, requiring varying degrees of retraining for each change. Some experts argue that people will alternate between work and "nonwork" on a more regular basis. For example, Fred Best uses Figure 3–8 to support his claim that, rather than a continuous period of work preceded by education and followed by retirement, workers will move in and out of full-time work, education, and leisure. This may fit well with the expressive values already described or with a pattern of recurring involuntary unemployment.

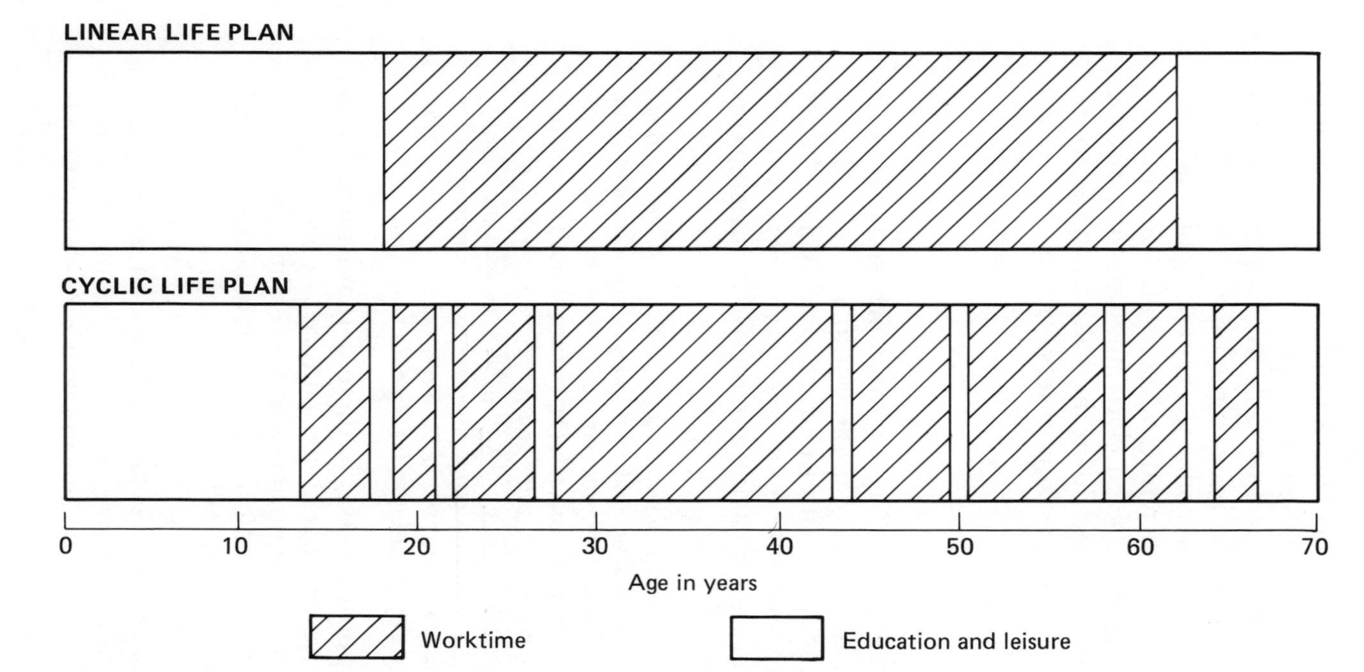

Linear Life Plan (the way life is now organized): An extended period of non-work at the beginning of life is followed by a solid period of work years and then another period of non-work. Under this plan, most increases in non-work are taken in the form of reduced workweeks and expansion of the time for education during youth and leisure during old age. Such expansion reduces the compressions of work into the mid-years of life but maintains the linear progression from school to work to retirement.

Cyclic Life Plan (the way life may be organized in the future): Non-work time is redistributed through the middle years of life to allow extended periods of leisure or education in mid-life.

Figure 3-8 Alternative Lifetime Patterns. Source: Fred Best, "Recycling People: Work-Sharing Through Flexible Life Scheduling," *The Futurist*, February 1978, p. 8. Used with permission.

Work at home will become a viable option for many, either by working out of their house or having an office but working at home for part of each week or month. For several years now, it has been argued that about 40 percent of what most white-collar workers do could be done at home, and more efficiently, without any electronic or computer equipment. Given electronic workstations, the difference between being down the hall, down the street, or in another town becomes less critical for many occupations.

Organizational Culture

"Organizational culture" is a term which summarizes a number of areas of change in the workplace; it is a principal condition for many of the trends in the workplace. James O'Toole concludes his book, *Making America Work,* with the following argument for paying attention to a changing corporate culture:[39]

> *Only changes in the philosophy and organization of work can overcome America's economic decline. And such changes can occur only when managers are willing to identify the values and assumptions that underlie the culture of their organizations, what the cultures* should *be. Only then will they see the need for change and be able to create in the work place the conditions of diversity, flexibility, choice, mobility, participation, security, and rights tied to responsibilities, which are necessary in making the culture of organizations congruent with the larger culture—conditions that would go a long way towards making America work again.*

Given this supposed need, is there a trend toward this more productive organizational culture? *In Search of Excellence* argues for its importance in successful companies, and *Business and Health Magazine* has a regular column on organizational culture. While there is little empirical research which supports the shift, and it is widely acknowledged that cultural change of any kind is difficult to achieve and maintain, there are likely to be interacting factors which will increase the importance of conscious efforts devoted to the development of a new corporate culture.

Pensions, Retirement Policies, and Social Security

Most people in the workforce today, except perhaps those under 30 to 35, assume that when they retire a combination of social security and private pension plans will provide them with financial security. Many are alarmed, however, for some data suggest that for many this is far too comfortable an assumption. According to the Trustees of the Social Security Trust Fund, under an "intermediate" set of assumptions combining some strong indices of economic growth with other more pessimistic forecasts, social security will be bankrupt by 2018. The key assumptions include 3.6 percent real annual GNP growth, 7.4 percent unemployment, inflation of 5.2 percent, and hospital inflation running at 12.4 percent. For the period between 1984 and 2008, these assumptions would result in an annual surplus of $11 billion for the combined Old Age, Survivor and Disability Insurance (OASDI) and the Hospital Insurance (HI) program. However, the HI trust fund faces the prospect of possible bankruptcy, and its mounting deficits could grow to such proportions that continued OASDI surpluses would be inadequate to keep the combined trust funds in the black beyond 2018.

Many critics contend that the assumptions behind the 2018 forecast are not likely to hold. Some argue that hospital inflation has reached its turning point and will not grow at anywhere near the rate forecast. Moreover, the hospital forecasts assume little change in the current illness patterns or current modes of treatment over the next three decades—a dubious assertion, given the trends in health care, particularly the prospects for compression of morbidity. These changes would aid social security solvency. On the down side, a growth rate of 3.6 percent is very high—almost competing with the 1960's. Likewise, 7.8

percent unemployment may well be low. Futurist Marvin Cetron argues that, given a healthy economy, the permanent unemployment rate is likely to become 8.5 percent because of the continual restructuring within the economy. Neither of these figures includes the potentially large number of jobs lost and not regained from automation and expert systems; and inflation is likely to run near 10 percent. This again is at least in part a result of the shift from a second-wave to a third-wave society, and also because the growing federal debt of the 1980's will affect the economy for years to come.

Some experts argue that the economic road for society in the next twenty-five years will not be an easy one. Many of the factors cited here—including the technological revolution, globalization of the economy, and the postulated economic decline—will, if they develop, block social security's already troublesome path by lowering the inflow from contributing workers. Thus, they say, bankruptcy for social security may come ten or even twenty years sooner than expected.

What is much less publicized is the uncertain state of many private pension plans. Many companies have failed to adequately fund or provide sound actuarial bases for pension plans established through collective bargaining or otherwise provided by the employer. This has led to the prospects for some retirees of a twenty-five year retirement period with unfunded or underfunded pension support. The recent dramatic growth in IRA's will give many future retirees a buffer against potential shortfalls from pension plans, or the social security system and will give them less incentive to want to maintain the social security system. For others, unable to adequately privatize their retirement, the expectations of a secure retirement based on social security and pension plans may become increasingly unrealistic. As Willis Goldbeck has argued:[40]

> *No degree of tinkering with the existing Medicare program will produce a solution to the pending deficit and close the program's coverage gaps for a population which is actively encouraging retirement periods of 20–*

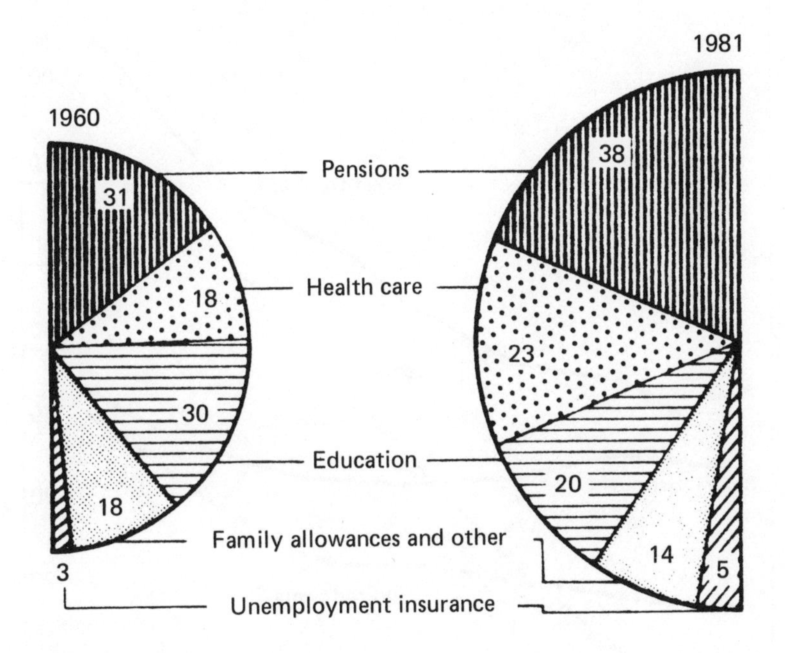

Figure 3-9 The Growth of Pensions in Relation to Other Social Expenditures. Source: *The Economist*, "Pensions After 2000," May 19, 1984, p. 59. Used with permission.

40 years. Neither Medicare, nor private medical insurance plans, nor Social Security were designed or financed to cover either the number of beneficiaries or the duration which tomorrow's demographics will demand.

These problems are not endemic to the United States. Throughout the major industrial countries, there has been dramatic growth of underfunded pensions. Figure 3–9 illustrates the growth of pensions in relation to other social

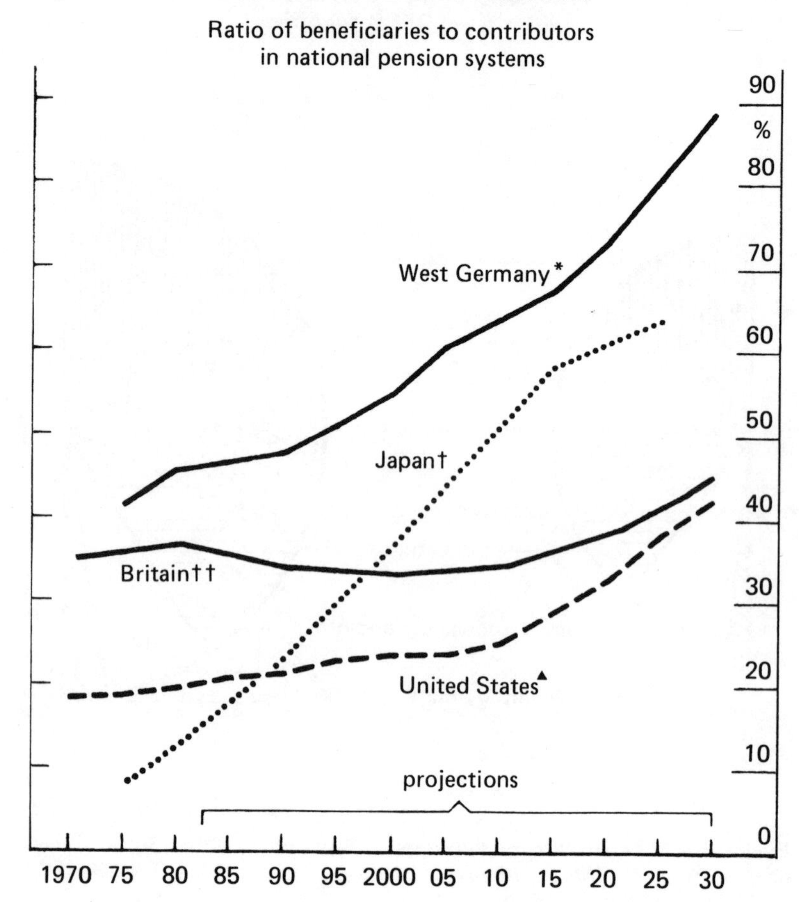

*Ratio of retirees to workers excluding public sector †Ratio of beneficiaries to contributors in the KNH ††Ratio of retirees to workers ▲Ratio of people aged 65 and over, to those aged 20-64

Sources: Foundation Nationale d'Economie Politique, Institute for Contemporary Studies

Figure 3-10 Ratio of Beneficiaries to Contributors in Four OECD Countries. Source: *The Economist*, "Pensions After 2000," May 19, 1984, p. 60. Used with permission.

expenditures from 1960 to 1981 in the seven largest OECD countries, of which the United States is a member. Figure 3-10 illustrates the ratio of beneficiaries to contributors in the national pension systems of West Germany, Japan, Great Britain, and the United States. The data for the United States are actually the ratio of those over 65 to those 20 to 64. The U.S. social security system is not tied to the financial earnings of the depositors, unlike the Japanese system, but instead is financed largely out of present collections and federal contributions. This creates long-term problems as the ratio of workers to beneficiaries changes. As one review put it, "the only real choices fall under two brutal options: lower spending, higher revenue."[41]

There are serious political difficulties with "lower spending," meaning a reduction of benefits, and "higher revenue," meaning larger payroll taxes. As this issue becomes more widely discussed, pressure for decisive action will invariably mount from a larger and larger current retirement-age population and from the baby boomers, who will realize that they may get relatively little from their decades of contributions to social security. The adjustments in retirement benefits may not only be in how much a retiree will receive, but when he or she can receive it. For the past four decades, the participation rates of older men in the workforce have declined, and projections are for this trend to continue through 1995.[42] However, the comparatively generous increase in payments to the elderly during the 1970's is not likely to be duplicated; a relative decline is more likely.

In the health area, the cutbacks in payments for the elderly are being looked at for cost savings. According to the Health Insurance Association of America (HIAA), the effect of federal action to curb costs for treatment of Medicare patients alone will "shift" $8.8 billion in Medicare bills to private insurance carriers and in turn to the employers who pay the bulk of the insurance premiums.[43] Further cost shifting—not just of the medical

Table 3-7. Improved Workplace Health Status Indicators

Work-related deaths	
1978	4,590
1979	4,950
1980	4,400
1981	4,370
1990	
Goal	3,750
Work-related injury rate (per 100 workers)	
1978	9.2
1979	9.2
1980	8.5
1981	8.1
1990	
Goal	8.3
Lost workdays rate (per 100 workers)	
1978	62.1
1979	66.2
1980	63.7
1981	60.4
1990	
Goal	55.0
Cases of skin disease	
1978	65,900
1979	67,900
1980	56,200
1990	
Goal	60,000

Source: U.S. Department of Health and Human Services, *Health-United States 1983 and Preventive Profile* (Washington: GPO), pp. 370–371.

care costs, but of other benefits as well—may cause some employers to refuse to offer, and still others to retrench on, benefit programs, especially for a steadily growing retirement-age population. At the same time, the pressure exerted by elders for protection against disability and disease and for income security will inevitably shape the public debate. This demand may lead, along with other forces, to a redefined role for the private employer in the

Table 3-8. The Ten Leading Work-Related Diseases and Injuries, United States, 1982

1. Occupational lung diseases: asbestosis, byssinosis, silicosis, coal workers' pneumoconiosis, lung cancer, occupational asthma	6. Disorders of reproduction infertility, spontaneous abortion, teratogenesis
2. Musculoskeletal injuries: disorders of the back, trunk, upper extremity, neck, lower extremity; traumatically induced Raynaud's phemonenon	7. Neurotoxic disorders: peripheral neuropathy, toxic encephalitis, psychoses, extreme personality changes (exposure-related)
3. Occupational cancers (other than lung): leukemia; mesothelioma; cancers of the bladder, nose, and liver	8. Noise-induced loss of hearing
4. Amputations, fractures, eye loss, lacerations, and traumatic deaths	9. Dermatologic conditions: dermatoses, burns (scaldings), chemical burns, contusions (abrasions)
5. Cardiovascular diseases: hypertension, coronary artery disease, acute myocardial infarction	10. Psychologic disorders: neuroses, personality disorders, alcoholism, drug dependency

*The conditions listed under each category are to be viewed as *selected examples*, not comprehensive definitions of the category.

Source: NIOSH, "Leading Work-Related Disease and Injuries—United States," *Morbidity and Mortality Weekly Report*, January 21, 1983.

purchase of a wide range of human services for employees of all ages or perhaps specifically to retirees and dependents.

Workplace Safety and Health

Arnold Brown and Edith Weiner argue that there are a variety of health-related trends in the workplace:[44]

> *... genetic testing to protect susceptible workers from hazardous substances, general improvement of the work environment to remove or diminish potential hazards (ranging from fluorescent light to electrical and magnetic fields, all of which may be carcinogenic), stress reduction, weaning from cigarettes, treatment for alcoholism, diet counseling, and mandatory vacations (for workaholics).*

The trend of corporate involvement in health care confirms the set of items that Brown and Weiner observe. Table 3-7 shows the trend in workplace health status in relation to the occupational health and safety goals for 1990 set by the Department of Health and Human Services. Recent trend data agree with Brown and Weiner. Table 3-7 notes that in two cases, worker-related injury rates and cases of skin disease, the 1990 goal had been achieved by 1981 and 1980, respectively.

Table 3-8 gives the leading work-related diseases and injuries. These trends will be altered by a variety of developments. For example, cardiovascular problems and some cancers will be affected by health promotion programs, such as increased screening both for specific genetic characteristics and health risk generally. In addition, workplace health will be improved through a variety of telematics devices which will facilitate screening and monitoring and allow the development of biochemically unique profiles for each worker, allowing much more precise determination of illness and wellness.

The cloud on the horizon is the potential for continuing bombardment of workers from toxic chemicals, particularly new chemicals whose effects are not immediately obvious, and from the higher risk of some of the fastest growing occupations. The movement to the information age carries with it some unique workplace hazards. Electronic and other high-tech manufacturing operations involve a host of toxic substances. A 1980 survey by the California Department of Industrial Relations found that the semiconductor industry had a worker illness rate three times that for general manufacturing. One of the new forms of "super" computer chips replaces silicon with gallium arsenide. Arsenic, a prime component of these super chips, is highly toxic, has been linked to skin and lung cancer, and can be lethal in large doses. It has been linked to at least one death in Boston's Route 128 high-technology area.

The semiconductor industry is just one example of new sources of toxic chemicals which may affect health in the workplace during the following decades. A report for the U.S. Environmental Protection Agency on the outlook for toxic substances notes a number of sources:[45]

> *The rapid expansion of biotechnology will create and release large amounts of proteinaceous biological materials and other intermediates and products. . . . High-technology materials and products such as composites, plastics, electronic chemicals, photovoltaics, and ceramics are being developed faster than their health and environmental effects can be assessed. These materials add new opportunity for chronic or accidental exposure to toxic substances during their manufacture, use and environmental dispersal, and final disposal.*

The extent to which these materials will affect health in the workplace is influenced both by the trend of increasing the variety and amounts of toxic substances and the trend toward paying more attention to the effects of these substances.

Occupational health and safety issues may grow in importance if current injury rates hold. In Table 3-3, we saw the fastest growing occupations over the next decade; those occupations with asterisks carry a greater than average risk of occupational injury or disease. Table 3-8 gives the ten leading work-related diseases and injuries in the United States. The injury incidence rates by industry division and employment size for 1977 and 1978 and for 1981 and 1982 are shown in Tables 3-9 and 3-10, respectively. Figure 3-11 graphically portrays the differing injury incidence rates by size of employment for 1978.

Occupational risks could increase in declining industries, when, in a fashion similar to that of the end of the sailboat era, companies try to squeeze out the last margins of production through riskier approaches. Also, economically marginal firms may spend less on the safety and health of workers. Earlier recognition of possible risks in the restructured workplace may become more important because of the regulatory problems introduced by the trend toward

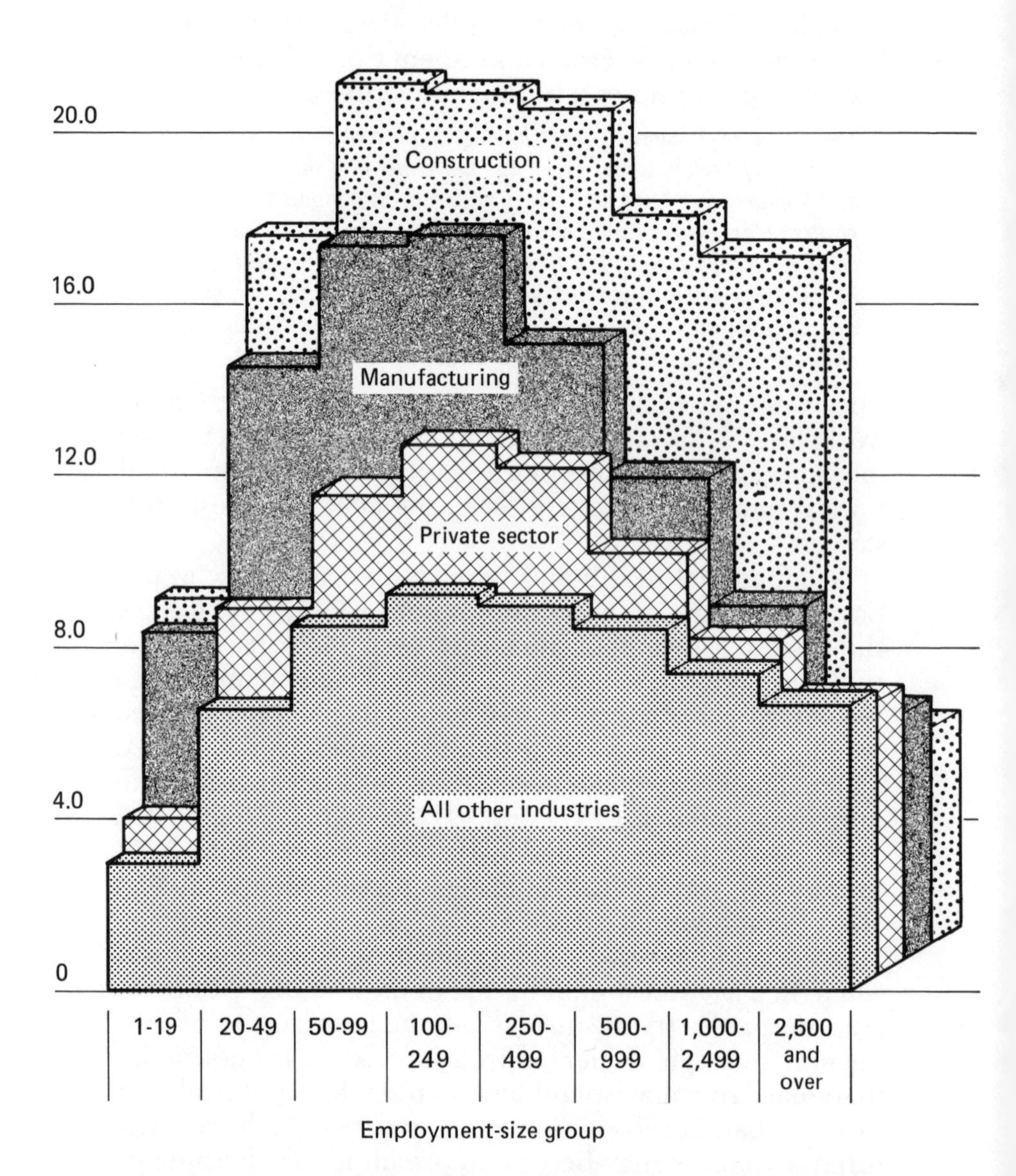

Figure 3-11 Injury Incidence Rates by Employment-Size Group, United States, 1978 (incidence rates per 100 full-time workers). Source: *Occupational Injuries and Illnesses in the United States by Industry, 1978,* U.S. Department of Labor, Bureau of Labor Statistics, August 1980, Bulletin 2078.

restructuring. As workers disperse to smaller worksites, promotion of occupational health will be much more difficult (for example, site visits will be more difficult and expensive), and new regulatory structures may be needed. There is likely to be an institutional lag as the regulatory system, created in response to large organizations, adjusts to the greater importance of smaller settings. For example, one evaluation of current research on health and safety issues estimates that 75 percent of the regulatory system's budget is targeted to mature or declining industries, while there is a great need to focus research on job sites in new, growing industries. It has also been pointed out that workers in small plants, especially those without labor unions, are less likely to participate in promoting health and safety issues; there are fewer workers to organize in support of such efforts, and employers may have more leverage in smaller units. Such relationships may embody the crucial aspect of changing corporate structures. The interplay between new structures and the attitudes of employers, workers, and the medical establishment is likely to determine the outcomes of future health problems in new worksites.

Employees as Human Capital

A number of trends are shaping the emerging perception of the employee as a source of capital. The following changes are among the more important trends:

- The shift toward "expressive" values, particularly as employees seek the means to express those values in the workplace.
- For a variety of reasons, more workers will seek increased flexibility in the workplace, including shared jobs, flextime, part-time, and a variety of sabbatical arrangements.
- Many younger workers, in particular, desire a much higher level of participation in the workplace decisions that affect them; they are more willing to follow rules if

Table 3–9. Occupational Injury Incidence Rates by Industry Division and Employment Size, United States, 1977 and 1978

Industry division	Incidence rates per 100 full-time workers[2]											
			Lost workday injuries				Lost workdays					
	Total cases[1]		Cases involving days away from work[3]		Cases involving days of restricted work activity only		Total lost workdays		Number of days away from work		Number of days of restricted work activity[4]	
	1977	*1978*	*1977*	*1978*	*1977*	*1978*	*1977*	*1978*	*1977*	*1978*	*1977*	*1978*
Private sector[5]	3.7	4.0	3.5	3.7	0.2	0.3	60.0	62.1	56.0	57.4	4.0	4.7
Agriculture, forestry, and fishing[5]	4.8	5.2	4.7	5.1	.1	.1	78.8	78.3	75.2	74.2	3.7	4.1
Mining	5.9	6.4	5.7	6.0	.2	.4	128.3	142.3	125.7	134.5	2.6	7.8
Construction	5.8	6.3	5.7	6.2	.1	.1	109.7	108.1	106.0	103.3	3.7	4.8
Manufacturing	4.9	5.4	4.5	4.9	.4	.5	79.3	82.3	72.3	74.2	7.0	8.1
Transportation and public utilities	5.2	5.7	4.8	5.2	.4	.4	95.0	101.3	86.9	93.1	8.1	8.3
Wholesale and retail trade	2.9	3.1	2.8	3.0	.1	.1	43.5	44.3	41.3	41.6	2.1	2.7
Wholesale trade	3.5	3.9	3.4	3.7	.1	.2	51.9	56.8	49.0	52.8	2.9	4.0
Retail trade	2.6	2.8	2.6	2.7	(6)	.1	40.0	39.1	38.2	37.0	1.8	2.1
Finance, insurance, and real estate	.8	.8	.8	.8	(6)	(6)	10.2	12.1	9.6	11.3	.6	.8
Services	2.2	2.3	2.1	2.2	(6)	.1	34.2	35.4	32.7	33.5	1.5	1.9

[1]In order to maintain the comparability of the 1978 survey data with the data published in previous years, a statistical method was developed for generating the 1978 estimates to represent the small nonfarm employers in low-risk industries which were not surveyed. The estimating procedure involved averaging the data reported by small employers for the 1975, 1976, and 1977 annual surveys.

[2]The incidence rates represent the number of injuries or lost workdays per 100 full-time workers and were calculated as: (N/EH) × 200,000 where,

N = number of injuries or lost workdays
EH = total hours worked by all employees during the calendar year
200,000 = base for 100 full-time equivalent workers (working 40 hours per week, 50 weeks per year).

[3]Also includes cases which involved both days away from work and days of restricted work activity.

[4]The number of days of restricted work activity include those resulting from cases involving restricted work activity only and days resulting from cases involving days away from work and days of restricted work activity.

[5]Excludes farms with fewer than 11 employees.

[6]Incidence rates less than .05.

Source: *Occupational Injuries and Illnesses in the United States by Industry, 1978*, U.S. Department of Labor, Bureau of Labor Statistics, August 1980, Bulletin 2078.

Table 3–10. Occupational Injury Incidence Rates by Industry Division and Employment Size, United States, 1981 and 1982

	Incidence rates per 100 full-time workers[1]							
Industry division	*1 to 19 employees*	*20 to 49 employees*	*50 to 99 employees*	*100 to 249 employees*	*250 to 499 employees*	*500 to 999 employees*	*1,000 to 2,499 employees*	*2,500 employees or more*
Private sector:[2]								
1981	3.5	7.7	10.9	11.6	10.9	9.3	7.5	6.0
1982	3.5	7.7	10.0	10.7	9.9	8.5	6.7	5.5
Agriculture, forestry, and fishing:[2]								
1981	6.3	10.5	13.8	15.6	16.1	14.6	29.9	17.7
1982	6.1	9.7	14.3	14.0	16.6	14.1	27.9	21.7
Mining:[3]								
1981	6.9	11.4	14.7	14.4	12.1	10.8	6.3	3.7
1982	6.4	12.5	13.6	12.8	9.4	8.8	4.1	2.9
Construction:								
1981	8.8	16.6	21.0	21.2	18.2	16.0	13.8	7.8
1982	8.4	16.8	19.6	22.1	18.0	16.5	9.7	7.7
Manufacturing:								
1981	7.6	13.0	15.9	15.4	13.1	10.4	7.6	5.4
1982	7.2	12.3	14.1	13.6	11.4	8.9	6.3	4.8
Transportation and public utilities:								
1981	5.2	10.2	11.9	9.0	8.9	8.9	8.0	9.1
1982	5.4	10.0	11.1	8.7	7.8	8.5	7.3	8.2
Wholesale and retail trade:								
1981	2.9	6.6	9.5	11.4	11.5	11.5	11.7	10.3
1982	3.0	7.0	9.4	10.5	11.5	11.1	10.6	10.4

Wholesale trade:								
1981	3.8	7.6	10.9	11.5	10.5	10.0	6.9	—
1982	3.9	7.5	9.9	10.4	8.8	8.8	7.0	—
Retail trade:								
1981	2.5	6.1	8.9	11.3	11.9	12.0	13.0	10.4
1982	2.6	6.7	9.2	10.6	12.6	11.9	11.3	10.4
Finance, insurance, and real estate:								
1981	1.2	1.4	2.0	2.5	2.7	2.6	2.2	2.1
1982	1.4	1.6	2.2	2.4	2.8	2.5	2.1	1.9
Services:								
1981	1.5	3.0	6.1	7.3	7.6	6.8	6.8	5.2
1982	1.6	3.0	5.5	7.2	7.1	7.0	6.7	5.4

[1] The incidence rates represent the number of injuries per 100 full-time workers and were calculated as: (N/EH) × 200,000, where
N = number of injuries
EH = total hours worked by all employees
200,000 = base for 100 full-time equivalent workers (working 40 hours per week, 50 weeks per year).

[2] Excludes farms with fewer than 11 employees.

[3] Employers engaged in surface mining or milling of stone, clay, colloidal phosphate, sand, and gravel, and independent construction contractors at surface areas of mines were exempted from enforcement of the Federal Mine Safety and Health Act of 1977 during the first half of calendar year 1982. Nonconstruction contractors were not affected by the exemption.

Note: Dashes indicate data that do not meet publication guidelines.

Source: *Occupational Injuries and Illnesses in the United States by Industry, 1982*, U.S. Department of Labor, Bureau of Labor Statistics, April 1984, Bulletin 2196.

they have had something to say about their formulation and are appropriately compensated for their inputs into the production process.

- As the discussion of values shows, many employees are placing a demonstrable value on the quality of products and services.
- The increasing perception by managers that since job retraining and job replacement costs are so high, high turnover rates are unacceptable—just as it is unacceptable to management to incur excessive costs due to poorly maintained facilities and equipment. Hence, there is a growing willingness to invest more in human resource development and training for workers.
- As the nature of work changes, resulting in greater numbers of employees in the knowledge and information sectors of the economy, such employees may demand more discretion in the exercise of their work, and hence more self-autonomy, in order to achieve optimal productivity.

These trends and changing perceptions, *in combination,* are causing many employers to view employees, or at least some of them, as a form of capital for which nurturing, maintenance, improvement, self-enhancement, and self-growth are both needs and opportunities. Economic growth is increasingly tied to the development of human capital, according to a report for The Council of State Planning Agencies:[46]

> *Human capital is the combination of innate talent, knowledge, skill, and experience that makes each human a valuable contributor to economic production. Learning is the process through which human capital grows. As we proceed through the transition to a new, postindustrial economy, human capital and the learning that generates it are becoming ever more critical to healthy economic development.*

Results of this growing management interest in human capital will include new emphasis on promotion of health, wellness, human resource development, training, and quality control. And, as the evidence about the positive relationship among worker participation, productivity, and quality increasingly becomes available, more companies will experiment with innovative employee-participation schemes, including equity participation.

The Future of Worker Organizations

Worker organizations—particularly unions, but professional associations as well—are facing a variety of contradictory trends. Labor union membership as a percentage of the U.S. workforce has been diminishing for some time. Union jobs in the manufacturing sector are among the ones lost to overseas production, so this aspect of the decline is likely to continue. Alternatively, the fastest growing areas of union membership include white-collar and professional and government service employees.

Our discussion about incentives, the elimination of traditional distinctions or levels within workforces, and the encouragement of risk-taking suggests that union activity in the years ahead will adjust to these new factors in their environment. A step in this direction is the work of the AFL–CIO's Committee on the Evolution of Work. Its 1983 and February 1985 statements recognize the changing environment and many of the trends reviewed in this book.[47,48] In response, the AFL–CIO committee recommends that unions develop new approaches to collective bargaining, that they address issues such as pay equity and worker participation in workplace decision-making processes, and that they develop union membership forms beyond solely workplace collective bargaining agreements.[49] Forty-one percent of union members are in white collar jobs; another 20 percent are craftsmen or foremen. From 1971 to 1983, membership in the AFL-CIO by public sector workers grew by over 1 million while membership in the private sector declined by 2 million. About 50 percent of full-time state and local government employees are organized, and the AFL–CIO committee argues that the unions will see a resurgence if they can meet the challenges of the changing work environment.[50] Management, on the other hand, is likely to continue to read the same trends and adjust working conditions in ways discussed above, in order to satisfy the changing needs and values of workers and thereby prevent union growth.[51]

Notes

1. Robert B. Reich, *The Next American Frontier* (New York: New York Times Books, 1983).
2. Amitai Etzioni, *An Immodest Agenda* (New York: McGraw-Hill, 1983).
3. Paul Hawken, *The Next Economy* (New York: Ballantine Books, 1983).
4. Eli Ginzberg, "The Mechanization of Work," *Scientific American,* September 1982, p. 67.
5. Donald N. Michael, *Cybernation: The Silent Conquest* (Santa Barbara, Calif.: Center for the Study of Democratic Institutions, 1962).
6. Marvin Cetron with Marcia Appel, *Jobs of the Future* (New York: McGraw-Hill, 1984), p. 21.
7. John Naisbitt, in interview, *Omni*, 1984.
8. American Council of Life Insurance, *The Changing Work Place: Perceptions, Reality,* Trend Analysis Program, ACLI, March 1984, p. 2.
9. Sam Cole and Ian Miles, *Stacking Up the Chips: The Distributional Impact of New Technologies* (London: Francis Pinter, 1985).
10. Arthur B. Shostak, "Options for Blue-Collar Workers," *Business and Health,* May 1984, pp. 13–15.
11. *Ibid.*
12. Farnsworth, "The Too-Mighty Dollar Takes a Toll of Manufacturing Jobs," *New York Times,* September 23, 1984, p. E3.
13. Susan Dentzer, "A Touch of 'Made in America,' " *Newsweek,* October 27, 1984, p. 102.
14. American Council of Life Insurance, *Forces in Motion: Identifying Potential Crisies,* Trend Analysis Program, ACLI, 1983, pp. 14–16.
15. Robert Beckman, *The Downwave: Surviving the Second Great Depression* (New York: E.P. Dutton, 1983).
16. Security Pacific National Bank, *Trends: The Once and Future Economy,* Futures Research Division, 1984.
17. Rosabeth Moss Kanter, *Change Masters* (New York: Simon & Schuster, 1983).
18. Jane Jacobs, *Cities and the Wealth of Nations: Principles of Economic Life* (Toronto: Random House, 1984).

19. David Morris, *Self-Reliant Cities: Energy and the Transformation of Urban America* (San Francisco: Sierra Club Books, 1982).
20. *Tarrytown Letter,* "Supermanaging in the 80s," p. 4.
21. United Way of America, *Scenarios: A Tool for Planning in Uncertain Times* (Alexandria Va: United Way of America, 1984), pp. 8–9.
22. Willis Harman, "Work," *Millennium: Glimpses into the 21st Century,* Alberto Villoldo and Ken Dychtwald, eds. (Boston: Houghton Mifflin, 1981).
23. James Robertson, "What Comes After Full Employment?" paper presented at The Other Economic Summit, 42 Warriner Gardens, London, England, SW11 4DU, June 6–10, 1984.
24. Wassily W. Leontief, "The Distribution of Work and Income," *Scientific American,* September 1982, p. 188.
25. Daniel Yankelovich, et al., *Work and Human Values: An International Report on Jobs in the 1980s and 1990s* (New York: Aspen Institute for Humanistic Studies, 1983).
26. *Ibid.*, pp. 47–50.
27. *Ibid.*, p. 53.
28. *Ibid.*.
29. James O'Toole, *Making America Work: Productivity and Responsibility* (New York: Continuum, 1981), pp. 43, 197.
30. Clement Bezold, "Lucas Aerospace: The Workers' Plan for Socially Useful Products," in *Anticipatory Democracy,* Clement Bezold, ed. (New York: Random House, 1978).
31. Harlan Cleveland, "The Twilight of Hierarchy: Speculations on the Informatization of Society," prepared for the Symposium on Information Technologies and Social Transformation for the National Academy of Engineering, October 4, 1984, pp.12–13.
32. Security Pacific National Bank, p. 4.
33. *Ibid.*
34. Vary T. Coates, "The Potential Impacts of Robotics," *The Futurist,* February 1983, p. 29.
35. Sam Cole and Ian Miles.
36. Arnold Brown and Edith Weiner, *Supermanaging* (New York: McGraw-Hill, 1984), pp. 49–52.
37. Gerard K. O'Neill, *The Technology Edge* (New York: Simon & Schuster, 1983).
38. Marvin Cetron with Marcia Appel, p.6.
39. James O'Toole, p. 187.
40. Willis Goldbeck, "Incentives: The Key to Reform of Medical Economics," Testimony to the House Ways and Means Committee, September 13, 1984.
41. *The Economist,* "Pensions After the Year 2000," May 19, 1984, p. 59.

42. *White House Conference on Aging,* U.S. Department of Health and Human Services, June 2, 1982.
43. Mary H. Cooper, "Healthcare: Pressure for Change," *Educational Research Reports,* Washington, D.C. Congressional Quarterly, August 10, 1984, p. 579.
44. Arnold Brown and Edith Weiner, p. 120.
45. Vary T. Coates, et al., *Toxics '95: The Outlook of Factors and Trends for Toxic Chemicals,* Office of Pesticides and Toxic Substances, U.S. Environmental Protection Agency, May 1984, p. 11.
46. Lewis J. Perelman, *The Learning Enterprise: Adult Learning Capital and Economic Development* (Washington, D.C.: The Council of State Planning Agencies, 1984), p. 1.
47. *"The Future of Work:* A Report by the AFL-CIO Committee on the Evolution of Work," August 1983.
48. "The Changing Situation of Workers and Their Unions: A Report by the AFL-CIO Committee on the Evolution of Work," February 1985. Washington D.C.
49. *Ibid.*
50. *Ibid.*, p. 12.
51. Don G. Keown "Personnel Trends in the '80's," *Office Administration and Automation,* December 1984, p. 60.

Chapter 4

KEY TRENDS SHAPING THE FUTURE OF HEALTH AND HEALTH CARE

Health institutions are in the midst of revolutionary change. Some of this change has stemmed from practical forces, such as cost and lack of consumer satisfaction. There are also some powerful conceptual and even philosophical shifts that fundamentally alter the way our society perceives health and health care. This chapter covers these trends in three areas: demographics, changes in the delivery and financing of care, and changing practices of health and health care.

Demographics

Aging and the Changing Burden of Illness

We have already discussed the forecasts for the aged in our population. Obviously, changes in the age distribution of our population are significant in any sector of our society. Forecasts make it clear that the proportion of the over-65 population to the under-65 population is expected to rise from around 20 percent in 1980 to 23 percent by 2000, 26 percent in 2010, 33 percent by 2020, and to 42 percent by 2030. By 2055, this figure could be more than 50 percent.[1] This forecast, moreover, assumes no significant change in life expectancy or retirement age.

The implications for trends in both health and health care are profound: There will be more elderly people with the potential for chronic illnesses, especially in the absence of health promotion programs and compression of morbidity. Further, the management of accelerating costs could become more and more difficult, with questions about the rationing of health care becoming increasingly unavoidable. The political difficulty of dealing with such questions will likely be complicated by the demographic trends of increased immigration and higher fertility rates among minorities. The needs of a younger, poorer segment of society may compete directly with greater demands for health care from the growing elderly population.

Already being called into question is the focus of medicine on the prevention of death and on the provision of heroic measures which slightly postpone the time of death. In the future that focus may be altered significantly. Nevertheless, geriatrics will become an increasingly important branch of medicine. A key issue is how geriatrics will develop: whether it will follow the traditional model of attempting to conquer death and disease, or whether it will focus more on the management of chronic diseases to postpone or prevent their manifestation until near or after death.

The role that employers play in providing benefits, including disease-prevention and health-promotion programs for retirees, may become prominent and may also undergo dramatic change. Some companies already have large percentages of retirees for whom they will be paying health benefits (often under a Medicare supplemental plan) for another ten to twenty years after retirement. There is growing interest now in getting the retirees and their families involved in health-promotion programs. Also, if expressive values dominate or economic hard times come, home care by family and friends may return to the prominence it had before pensions, social security, and publicly funded medical care became widespread and commonly accepted. Simultaneously, nursing homes and nursing-home chains will be affected by the trends toward self-reliance and participation in health care. As a result, they will probably become more hospitable toward residents and involve their residents more in self-management.

Changes in the Delivery and Financing of Care

Health Care Expenditures

Health care expenditures have risen dramatically over the last two decades, from about 6 percent of the GNP when Medicare and Medicaid were approved in 1964 to nearly 11 percent in 1984. Some background on expenditures is relevant:[2]

> *In 1982, health care expenditures in the United States totaled $322.4 billion, an average of $1,365 per person, and comprised 10.5 percent of the gross national product. . . . Hospital care expenditures continue to claim the largest share of the health care dollar, accounting for 42 percent of personal health care expenditures in 1982. Physician services, dentist services, and nursing home care account for 19 percent, 6 percent, and 9 percent respectively.*
>
> *In 1980, females represented 52 percent of the population but accounted for 58 percent of personal health care expenditures. Although the elderly represented only 11 percent of the population, they accounted for 31 percent of personal health care expenditures.*
>
> *Diseases of the circulatory system accounted for the highest amount of personal health care expenditures ($33 billion), followed closely by diseases of the digestive system ($32 billion including $15 billion for dental care). Hospital care usually accounted for the largest share of expenditures in each disease class. However, the proportion varied from about 35 percent for diseases of the nervous system and sense organs and diseases of the digestive system to more than 60 percent for neoplasms, injury and poisoning, and mental disorders.*
>
> *The 1.3 million Medicare enrollees who died during 1978 comprised 5 percent of Medicare enrollment but accounted for 28 percent of program expenditures.*

Health care expenditures are at their current level because of a variety of historical forces, primarily changes in lifestyle (richer diets and less exercise, and diminution of family and community support networks,) changes in medical care delivery (which encouraged high-technology care, delivered by specialized providers, primarily institutions) and changes in financing (insurance programs which removed incentives from providers and consumers to contain cost or identify and measure quality). All of these historical forces are being modified by the trends in the delivery and financing of health care. The overall effect of the conflicts among the trends led the Institute for Alternative Futures to forecast a range of from 6 to 13 percent of GNP going to health care by 2010.[3]

Telematics (Emerging Computer and Communications Technologies)

A host of health care technologies will be put to use over the next twenty-five years, and among the most profound in their impact will be information technologies. The development of the information society may have its most profound and immediate applications in the area of health care. Health care is essentially the delivery of information on treating illness with various degrees of personal human care. Health care will be dramatically affected by a host of information and communications devices. This field of "telematics" will include diagnostic software; more sophisticated medical record keeping; the ability to communicate medical knowledge and patient observations virtually anywhere; and the ability to develop personal, biochemically unique definitions of illness, health, and wellness.

In the software category, "artificial intelligence" or "expert system" programs are being developed throughout the United States. Ultimately artificial intelligence technology may include devices that mimic much of the reasoning capacity of the human brain. Although the last ten years have seen dramatic advances, they have been limited by the current structure of computer architecture, which the fifth-generation computer may overcome. Currently there are diagnostic software programs, like CADECEUS, which can diagnose specific health problems (in CADECEUS's case, stomach problems) better than a general practitioner and can compete with the best specialists of the field. In the years ahead diagnostic and prescriptive software will improve, not only for specific fields but for most areas of medicine. Simultaneously the space requirements will diminish considerably for storing the software programs, the knowledge bases, and the personal histories of individuals.

Medical record keeping will become much more sophisticated. Very effective patient record and medical care management systems, such as Larry Weed's PROMIS, have been slow in gaining use in medical centers, but the changing incentives of the cost-management efforts already described are rapidly making health care providers seek out better analytical software, not only for cost management but for treatment. Combined with this will be an increased capability to monitor the outcomes of care.

A more familiar type of computer assessment program is the health-risk appraisals developed by a variety of firms. Whole packages are being developed and marketed to corporations for considering the full range of health and health care of their employees. Several leading conceptualizers in the health care field are developing information products that can be used for monitoring health and health care and for facilitating health promotion programs.[4-6] These programs, now focused on corporations, are also being developed for consumers interested in self-care.

An important outcome of these various programs and products will be the development of health care that deals with the biochemical uniqueness of individuals. Current definitions of illness based on clinical tests, for example, indicate that a person is ill if his or her test results are greater than two standard deviations from an epidemiologically derived norm for the nation or the region. Yet the uniqueness of each of our bodily chemical factories is as great as the uniqueness of our fingerprints. The new devices (particularly home health care equipment) allow for easy, effective, nearly continuous monitoring of physical conditions and moods/mental states. These devices will allow individuals to develop their own personal definitions of health and illness.

The individual's capacity to understand his or her unique biological profile may be significantly enhanced by forecasted technologies. For example, the combination of better diagnostic and prescriptive software with personal monitoring devices, new forms of drug delivery, and bioelectrical therapy may lead to the "hospital-on-the-wrist" by the turn of the century, if not before.[7] This technology would allow individuals not only to monitor unique characteristics but also to choose the philosophy of health care and medicine on which their "hospital-on-the-wrist" would be based. They could also determine the degree of self-care or type of therapy they want—for example, high-tech medicine or high-therapeutic touch medicine.

Related developments in artificial intelligence may also lead to more active consumer monitoring of health care providers. Telematic devices may allow consumer groups to collect and disseminate information and evaluations on therapies as well as on providers. Similarly, clinical trials for new drugs or other therapies will be aided by the increasing sophistication and decreasing cost of information technology.[8,9]

The growing capability to handle medical record information may converge with the increasing knowledge of genetic characteristics and predispositions. It is not unlikely that medical record systems in the future will include the coding of an individual's basic gene structure. This genetic screening, used first to identify the sensitivity of workers to workplace hazards, will have far broader applications in the twenty-first century.

There are a host of issues involved with forecasts for this equipment. Privacy, accuracy, and liability questions immediately arise and will have to be dealt with. The access to these devices will be greatest for the affluent and for those whose employers acquire them. Ultimately, however, they are likely to become as ubiquitous as television.

Biomedical Research Breakthroughs

Biomedical research is likely to yield significant breakthroughs between now and the first years of the twenty-first century; these advances are so significant, Lewis Thomas argues, that we are on the verge of a disease-free society.[10] Pharmaceutical R&D is likely to yield very specific diagnostic tools as well as vaccines for many cancers and possibly heart disease. Also, much more effective and better-targeted cures for cancer, heart disease, and other major diseases are on the horizon. Antiviral agents and the capacity to control autoimmune diseases will emerge as well.

Some researchers feel that the major diseases are amenable to prevention or postponement through lifestyle changes (see the discussion on compression of morbidity later in this chapter). Not all researchers agree with these forecasts. Others feel that it is a romantic hope to expect either research breakthroughs or lifestyle changes to have the dramatic results suggested by Lewis Thomas. Forecasting in this area is made more difficult because of the role which serendipity plays in all invention, but particularly biomedical research.

Yet in this area in particular it is relevant to recall Clarke's First Law:[11]

> *When a distinguished but elderly scientist states that something is possible, he is almost cerainly right. When he states that something is impossible, he is very probably wrong.*

This law, and much research experience, suggests that some experts may be timid in forecasting changes in their field. The implication is that change may always be more dramatic than we can predict. On the other hand, Clarke's First Law also tells us that we can expect many of the predictions of established scientists to be realized. Many scientists argue that recent developments in biotechnology immunology, and molecular biology constitute a revolution.[12] The following areas for research breakthroughs have been identified as major possibilities by eminent scientists, including a number of Nobel laureates.

Immunology. The last decade has seen tremendous growth in our knowledge of the immunological system. It is possible that continued progress could lead as far as immunotherapy to stop the diseases of aging.[13] The diseases for which many believe vaccines may be developed by the year 2000 include hepatitis A and B, chicken pox, herpes simplex, meningitis, gonorrhea, tooth decay, and peridontal disease. Many others recognize a real possibility that vaccines also may be developed for heart disease and for cancers.[14] Understanding the immune system is also expected to lead to effective interventions in such autoimmune diseases as rheumatoid arthritis and lupus erythematosis by the end of this century.[15]

Brain Research and Neurotransmitters. Brain research is at the exciting point where a number of major developments may converge. Studies of immunology, neurobiology, and endocrinology are achieving new understanding about how the brain functions and how it affects physiological changes. Biotechnology now provides sophisticated research tools, such as monoclonal antibodies, allowing new studies of neurotransmitters and receptors that show cellular development within the brain. Knowledge is rapidly moving even beyond the cellular level to subcellular studies through which the role of specific molecules may soon be understood.[16]

Another set of tools, the PET, CAT, and MNR scanners, have already linked the computer revolution and medicine. They are likely to have a significant impact on both research and clinical management of neurological diseases. Illnesses such as Alzheimer's disease and Parkinson's disease have been among the most difficult to treat and the most heartbreaking to live with. Recent research on neurotransmitters has indicated possible treatable deficiencies, which may lead to cures for these diseases. Also, as the specificity of various neurotransmitters is better understood, a host of new tranquilizers and analgesics may be developed. They can be expected to have fewer side effects than today's best psychotropic drugs, and, according to Roy Vagelos, director of planning and research for Merck Sharp and Dohme, such drugs may be developed by 1990.[17] Another area in which the work on neurotransmitters offers great hope is in the treatment of mental illness, particularly schizophrenia, which has been linked to imbalances of various substances in the brain.[18]

Basic Genetic Knowledge and Genetic Engineering. Clones of human insulin and interferons are already familiar products of the biotechnology revolution. Soon, other products such as the anticlotting substance Factor VIII, human growth hormones, and other peptides will become available in greater quantities through genetic engineering. It is also likely that within this decade we will see the correction of single-gene-defect diseases—primarily the blood-forming ones—where the affected cells can be removed from the body, treated in vitro, and returned.[19]

DNA manipulation is also leading to new diagnostic techniques and products. Many of these developments are soon expected to reach the market, while others have exciting long-range potential. DNA probes are being designed for a variety of bacteria and viruses, allowing quick identification of infections as well as determining the drug resistence of the pathogens.[20] Besides infections, the diagnostic techniques may be important to research into diseases such as arthritis, cystic fibrosis, dystonia, neurofibromatosis, Wolf's syndrome, and some forms of muscular dystrophy and cancer.[21]

Monoclonal Antibodies. Research on monoclonal antibodies may provide new and more targeted cancer therapy than has been possible to date. This may prove particularly important for the treatment of cancers in which less well targeted therapies have created drastic side effects. Also, monoclonal antibodies—sometimes called DNA probes, as discussed in the preceding section—are likely to revolutionize diagnosis. They will allow more sensitive, simple, and inexpensive tests which may be used at home or in doctors' offices rather than in hospitals or certified laboratories. These tests may help promulgate a system of self-care if they become widely used by people for self-diagnosis of diseases susceptible to prevention.

New Delivery Systems. A variety of new delivery systems will become common in the years ahead for a wide range of drugs, including nasal sprays, implantable pumps that provide as-needed release of chemicals, controlled release polymers, and the monoclonal antibodies used as precisely guided missiles in the body.[22]

"Soft Technologies." In addition to the above areas for research breakthroughs in areas such as immunology, neurotransmitters, heart disease, and cancer, there is another area consistent with the self-care, promotion of health, use of the mind in healing, and other trends. They constitute what can be called "soft technologies." The variety of personal growth techniques will give persons a high degree of "body wisdom" that will for many, for example, allow for control of autonomic function and self-enhancement of the immune system.

These soft technologies may well join with cancer vaccines as the "fully decisive" technolgies which Lewis Thomas argues will prevent or cure disease. This could contribute to the compression of morbidity, with more workers able to continue working later in life, and lead to dramatic changes in the nature of employment in the health care industry. The biomedical breakthroughs will lead to a variety of more effective diagnostic and treatment capabilities. In the best of all possibilities, by 2010 preventive vaccines or definitive cures will have been found for most cancers, heart disease, and many of the conditions of aging, while "soft technologies" will also have reduced their incidence and have given more effective therapies.

Organizational/Institutional Change

It is not too much to say that the medical care system, in terms of its organization, delivery, and financing mechanisms, will change more in the next five to ten years than it has changed in the last fifty years. This change is associated with at least three sets of forces. First, there is widespread recognition that change is necessary. Second, new concepts about health are developing, making fundamental change conceivable. The third force, the challenge to well-defined traditional relationships as change occurs, will be discussed under growth of investor-owned organizations.

The most generally recognized force is that a critical mass of opinion has formed among key "publics"—including business groups, consumers, and legislators—that change is necessary. The rising cost of medicine during a period of economic instability has led many to reexamine assumptions about the role of health care in health. Efforts to control public and private expenditures for health care have challenged not only public policy but also the health care system itself. The most visible public issues have dealt with expenditures for services and access to care, particularly for the poor and the elderly. Increasingly, however, those who pay for medicine are asking questions about what they are buying and the availability of less costly options. These questions are encouraged by the interplay between the other two sets of forces promoting change in the delivery and financing of care.

The second set of forces is created by major conceptual changes that have taken place and are taking place in the way in which we think about our health. These profound conceptual shifts include those related to notions of individual control over health status, holistic health, and promotion of health and wellness programs. In the face of the changing concepts about health and health care, new questions have been raised about the efficacy, effectiveness, and efficiency (or cost effectiveness) of various interventions. As pressures mount from the rising cost of health care, more questions will be asked about the scale of effort and the trade-offs between expenditures for prevention and for cure. Simultaneously, new forms of organization and competition are developing among providers of care, opening new possibilities for change.

Diversification

Diversification is underway in delivery through the *alternative delivery systems* such as health maintenance organizations (HMO's) and preferred provider organizations (PPO's) and more directly through the development of *alternative delivery sites*—for example, ambulatory surgery centers, urgent care centers, birthing centers, home health care, retail health and medical care centers, and hospices. HMO's have grown in enrollment from under 3 million in 1970 to over 12 million in 1983, and are projected to enroll over 30 million people, 15 percent of the population, by 1990.[23] PPO's have developed rapidly, from 107 in 1983 to as many as 1,000 at the beginning of 1984.[24] The introduction of HMO's has shifted the incentives to reduce costs to the HMO itself because of the fixed yearly fee per person. The PPO's negotiate a price for services but don't absorb the risk. However, the employers or insurers who are seeking PPO arrangements are becoming much more directly involved in health care management.

Ambulatory surgery centers increased from 55 units in 1975 to 200 in 1983; growth is projected to be at the rate of 25 to 30 percent per year with a market potential from $1.2 to $1.7 billion.[25] In addition, the number of urgent care centers is expected to grow by 50 percent per year, with patient volume growing by 20.4 percent per year, and a market potential of $250 to $275 million. Retail health and medical care service centers, either freestanding or in shopping malls, are also expected to have large growth, with a market potential of $3.7 billion.[26] Finally, home health care is expected to have significant growth, projected at 13 to 20 percent per year, with growth to $7 to $9 billion by 1986,[27] and to $18 billion by 1990.[28]

Aggregation Among Providers

The growth of consolidations, associations, mergers, and a variety of other means of aggregation among health care providers, particularly hospitals, stems from a number of factors: access to capital, to superior management skills, to large-volume purchasing discounts, and to increased marketing power. These aggregations include new educational associations, for example, the Association of Western Hospitals; multihospital associations, such as Intermountain Health Care; and chains of for-profit and nonprofit hospitals, as well as systems of emergency care, surgicare chains, and so forth.

Today, one out of three hospitals is affiliated with a multihospital system, and estimates are that in five years it will be one of every two hospitals.[29] Also, while one out of two physicians practices in a group setting today, it is estimated that this figure will increase to 9 out of 10 by 2000.[30] These consolidations are likely to affect medical practice in a number of ways. More expensive and effective diagnostic equipment and computerized medical record systems will be available to doctors who can aggregate capital. The growth of alternative providers may be affected if group practices employ less-expensive providers who might otherwise create competition.

Growth of Investor-Owned Health Organizations

The growth of investor-owned organizations for care, such as Hospital Corporation of American (HCA), Humana, and American Medical International (AMI), will have an increasingly powerful impact on health care delivery. These three are primarily hospital companies, but such firms also frequently include chains of nursing homes, pharmacies, home health care providers, diagnostic centers, urgent care centers, and blood centers.

This investor-owned growth has occurred largely because of the high volume and relatively high margins for return in health care, along with the relatively poor management skills available in some parts of the voluntary sector in the health care system. Given the variety of trends considered here, the profitability of providing health care for the bulk of the population may be significantly reduced over the next twenty-five years. This is particularly true if compression of morbidity is achieved for many through promotion of health or definitive cures or vaccines for cancer and heart disease. Also, some of the management advantage of the chains may be lost as these management approaches become more readily copied and as computer software for organizational management becomes more accessible and less expensive. (Alternatively, it will be the well-financed institutions—not always for-profit—that can get the most rapid access to the new equipment.)

Cutting across all of these organizational developments is the third set of forces: a profound and deep change in roles and power relationships in the health care system. In the next few years, the biggest losers will be physicians, particularly those physicians not associated with one or more of the larger, more competitive health care organizations or who do not have an already-established group of patients with flexible insurance. The second largest group of losers will be small hospitals, particularly community hospitals that cannot take advantage of the lower management overhead of chains or associations and which do not diversify quickly enough. The third largest group of losers will be hospital administrators who have not recognized that they are conducting businesses that are vulnerable to market forces. Those who will gain as the system evolves will be managers skilled in developing new products and services. Those who can position their businesses to take advantage of change and those who are skilled in marketing are likely to do well. Also, those employers who recognize the extraordinary leverage they possess as large purchasers of care will gain competitive advantages.

Changes in the Number of Physicians

The number of active physicians increased 40 percent from 1970 to 1980 and is projected to increase by over 50 percent between 1980 and 2000. The number of active physicians per 100,000 people, which rose from 152 in 1970 to 191 in

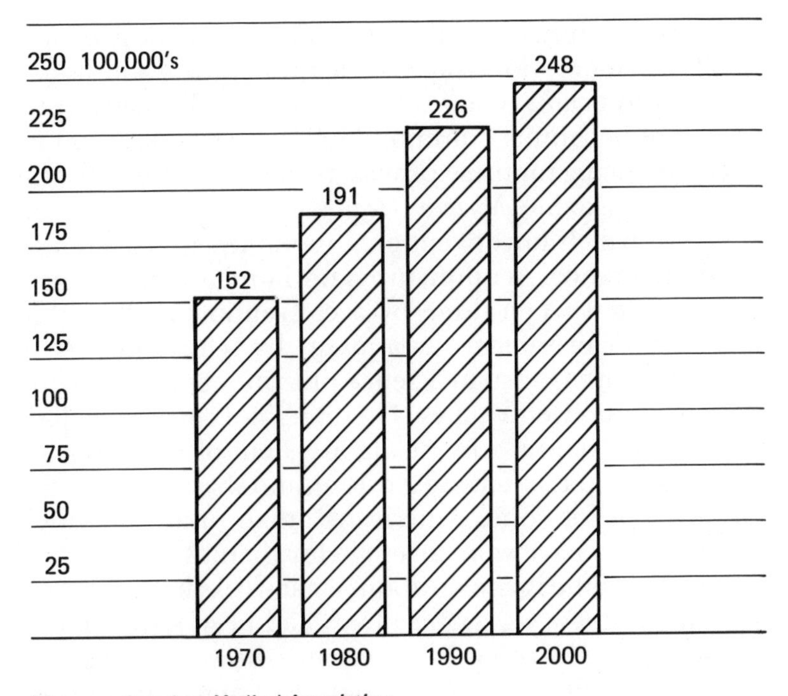

Figure 4-1 Ratio of Active Physicians to Population, 1970-2000. Source: *The Restructuring Health Industry: Progress Through Partnerships* (Minneapolis: The Health Central System, 1984), p. 33. Used with permission.

1980, is forecast to be 248 by the year 2000 (see Figure 4-1). By 1990 half of all physicians will have begun practice since 1978. This growth will have fostered competition among physicians while weakening their capacity to organize against other interests in the health care system.

Growth of Alternative Providers and Alternative Therapies

There has been a large growth of alternative providers and therapies in recent years, both within the conventional medical hierarchy—for example nurse-midwives, nurse-practitioners, nurse-anesthetists—and among nonconventional providers—for example, acupuncturists and homeopathic physicians. Guides to the field of alternative providers are becoming common.[31,32,33] A critical question involves financing for these providers. A recent review for the hospital magazine *Modern Health Care* reported on the degree to which unconventional therapies are covered by health insurance. Acupuncture and homeopathy are generally covered when done by a conventional (allopathic) physician; chiropractic treatment by chiropractors is covered, as are Christian Science practitioners and naturopaths in about twenty states. Therapies not covered include acupressure, naprapathy (using a system of manually applied movements to release tension), and shaman/medicine man (except for some American Indian union members whose unions have won coverage for them).[34]

Relatively little is known about what works in health care. Until the last few years, incentives had been structured to discourage our learning. Recent trends, however, have facilitated the monitoring and comparing of outcomes of competing therapies. These trends are in medical record keeping, which is becoming computerized; treatment monitoring devices and software, which are rapidly developing; and, employer, insurer and consumer interest in receiving accurate information on what works. These developments will expand the range of choices for workplace health care and promotion of health.

The pressure from a wide range of conventional practitioners, such as nurses and alternative providers, for greater legitimacy and economic rights will have an impact, long before 2010, on the next trend topic: licensure of physicians.

Questioning of Licensure

Licensure is a social invention, a public policy that grants physicians a monopoly on the right to practice medicine.

Other health care providers are licensed but provide little direct competition to physicians (dentists, podiatrists, and pharmacists, for example). Licensure laws established in the early part of this century sought to protect consumers from quacks and fraud in an area where consumers could not know enough to protect themselves. Dissatisfaction with physicians, competition from alternative providers, and the increasing sophistication of consumers have led to proposals to replace the medical licensing system that has been in place since the 1920's with a certification system that opens up competition between alternative providers of health care. For example, in 1983 a proposal nearly passed the California medical licensing and disciplinary board to amend the licensing laws to allow a wide assortment of health practitioners to maintain independent practices in order to provide many services that are now within physicians' legal domain. Recently, a Texas court declared the Texas medical licensing law unconstitutional on the grounds that it interfered with people's rights to seek health care from nonphysician practitioners.[35]

Between 1985 and 2010 this trend will accelerate through the expansion of the number and type of providers, the state of telematics in health care, and more assertive consumer monitoring. Public policy tasks will focus increasingly on the systems by which we develop "health outcome measures" and the effectiveness of various monitoring programs in the marketplace, run by both providers and consumers.

Changing Practices of Health and Health Care

Practical and conceptual change is a major force reshaping health care. A number of aspects of this change are visible today, affecting the behavior of practitioners, payers, and consumers of health care.

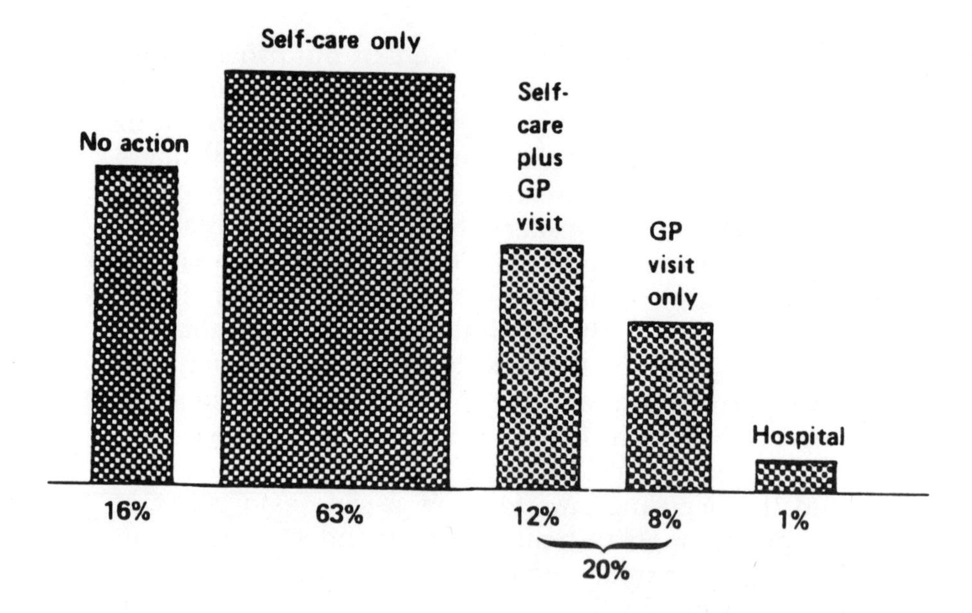

Figure 4-2 What People Did About Their Symptoms, 1978. Source: Arthur C. Hastings, James Fadiman, and James S. Gordon, *Health for the Whole Person* (Boulder, Co.: Westview, 1980), p. 88. Used with permission.

Self-care

More self-care has always been provided than professional care, whether the source has been individuals themselves or their family, friends, or neighbors. For example, a 1978 study reported by Arthur Hastings, James Fadiman, and James Gordon in their book, *Health for the Whole Person*, showed that 75 percent of the time some form of self-care was used, while a doctor visit occurred only 20 percent of the time (see Figure 4–2).[36] Figure 4–3 from a similar survey taken in 1983 showed that 60 percent treat themselves (even though 11 percent of this self-care is with a prescription drug at home), and only 9 percent go to a doctor or a dentist.

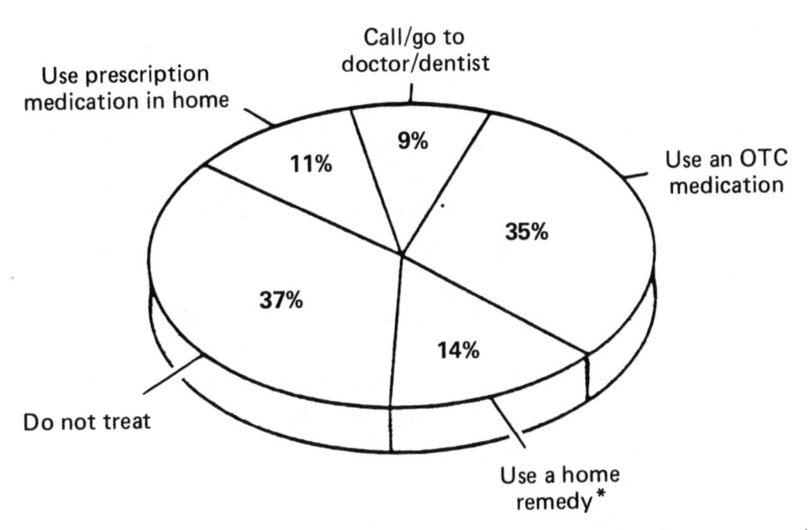

(Total adds to 106% because more than one action was taken in some cases.)

*Home remedies include salt-water gargles for sore throat and baking soda paste for bee stings—not medicines.

Figure 4-3 What People Did About Illnesses, 1983. Source: Harry Heller Research Corp., *Health Care Practices and Perceptions: A Consumer Survey of Self-Medication* (Washington, D.C.: The Proprietary Association, 1984). Used with permission.

There has, however, been a renewed interest in self-help as the promotion of health and wellness movements have focused on the role that the individual must necessarily play in health and healing. The growth of the home diagnostics market attests to the public's interest in personal involvement in health. Information resources have developed from a variety of sources. Physicians such as James Fries and Donald Vickery have developed self-care guides. It will not be long before the information in printed guides such as these, linked with body-function monitoring equipment, is a common technology for most households. Consumer and patient groups have created sophisticated medical library and research facilities, such as Planetree in San Francisco where not only hospital patients but others in San Francisco or around the United States can get the benefit of advanced medical journal knowledge by mail or phone. The American Red Cross, following a major self-study, has begun to aggressively expand its courses beyond first aid and CPR to broader self-care and health-promotion topics. *Medical SelfCare* magazine, edited by physician Tom Ferguson, has become a major catalogue of evaluations of treatments and resources for effective self-care. Several best-selling books, from the *People's Pharmacy* to books on diet, exercise, and life extension, attest to the growing concern with self-care. The market for self-care is likely to continue to grow as the equipment available, at first to upscale buyers, becomes more sophisticated.

The Promotion of Health and Wellness Movements

Survey data, and even more powerful trend analysis data such as those elicited by the Naisbitt group in its *Trend Report Publications*, show an important and arguably irreversible trend toward the promotion of good health in our population. The trend is particularly visible in the growing number of workplace health promotion programs. Over the last decade there has been explosive growth in the number of hospitals adopting wellness programs, and prospective payment can be expected to increase the number of hospitals offering wellness services in their communities.[37] A number of corporations are also marketing their programs to promote health, with some being franchised. People in the field are predicting tremendous growth in the near-term future.[38]

While the relationship between health promotion programs and longevity has remained mostly hypothetical, some recent studies offer important evidence of beneficial results of more healthful lifestyles. Research by Paffenbarger and Associates at Harvard and Myers Friedman, et al., in San Francisco, strongly suggests that regular fitness activity increases life expectancy. When combined with a variety of targeted stress-reduction programs and regular fitness activities, exercise reduces the likelihood of further heart disease in patients having already experienced heart failure. Animal studies also have shown that exercise can delay the development of atherosclerosis.[39]

Although health-promoting activities have been widespread in recent years, much of the program development activity has taken place at the worksite or at other sites sponsored by the employers. Most major U.S. corporations either have developed health promotion programs or have them seriously under consideration. Both the Washington Business Group on Health, a membership organization based in Washington, D.C., providing over 200 organizations with health affairs information, and the University of California Medical School at San Francisco, which is monitoring programs by companies in California (under the leadership of Ken Pelletier, author of *Unhealthy People, Unhealthy Places*),[40] report steadily increasing program development and activity.[41] While studies are being conducted and evaluations undertaken, the jury is out on many of the long-term effects of such programs on the health of employees and on the medical care and disability costs to employers.

Many short-term results, however, are positive with respect to moderating such costs, and almost all research to date demonstrates that employees who regularly participate in health and fitness programs are more productive than those who do not. Employees who exercise also express more satisfaction with their jobs. One survey of corporate fitness programs, for example, showed that they led to 1.5 percent to 15 percent reduction in employee turnover, which meant a cost savings between $4,000 and $8,000 per employee due to the decreased training costs. Other programs have claimed impressive declines in leaves of absence and disability insurance payments.[42] The largest impact of

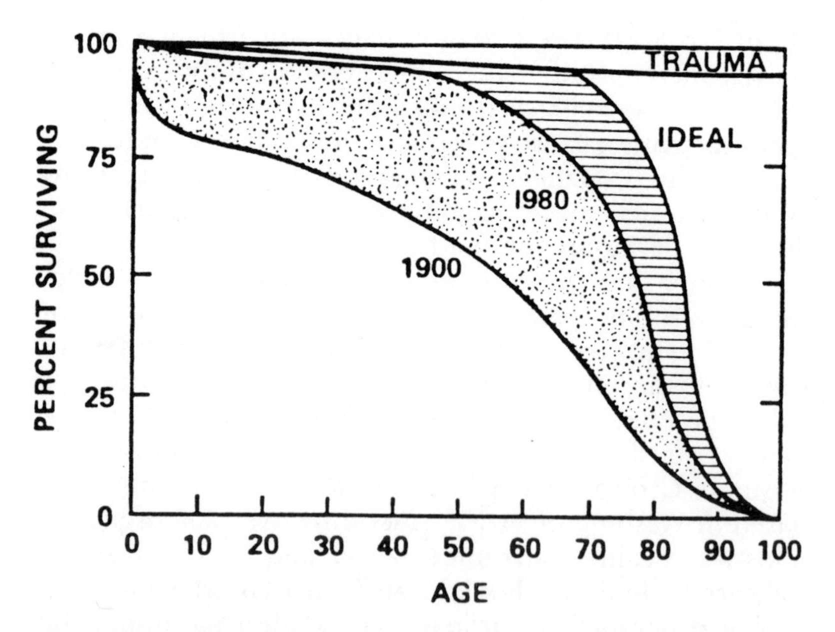

About 80 per cent (stippled area) of the difference between the 1900 curve and the ideal curve (stippled area plus hatched area) had been eliminated by 1980. Trauma is now the dominant cause of death in early life.

Figure 4-4 The Increasingly Rectangular Survival Curve. Source: James F. Fries, "Aging, Natural Death, and the Compression of Morbidity," reprinted by permission of *New England Journal of Medicine*, July 17, 1980, p. 130.

health promotion and wellness movements, if results are as favorable as expected, will be on the compression of morbidity.

Compression of Morbidity

James Fries, in *The New England Journal of Medicine*, his book *Vitality and Aging*, and elsewhere, has proposed a theoretical framework to explain evidence that more people are living

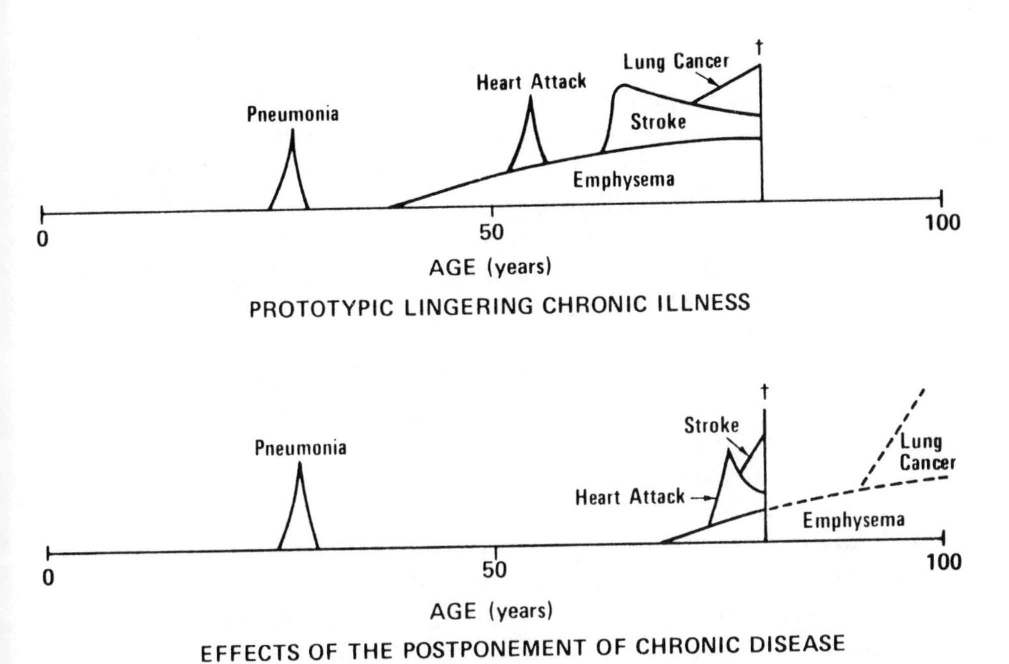

The compression of morbidity. The ability to postpone chronic disease, taken together with the biological limit represented by the life span, results in the ability to shorten the period between the clinical onset of chronic disease and the end of life. Infirmity (morbidity) is compressed into a shorter and shorter period near the end of the life span.

Figure 4-5 Effects of the Postponement of Chronic Disease. Source: J.F. Fries and L.M. Crapo, *Vitality and Aging* (San Francisco: W.H. Freeman, 1981), p. 92. Used with permission.

longer, healthier lives.[43,44] Fries argues that there is a fixed human life span (the biological limit to the length of human life) averaging somewhere between 85 and 100 years. Life expectancy (the number of years a person or group can be expected to achieve at birth) is increasing in the United States, as shown in Figure 4–4.

The major health problems which keep life expectancy below the life span are now the chronic diseases. But in many if not most cases, the onset of these diseases can be delayed through health promoting behavior and disease prevention. The ideal strategy thus becomes to compress morbidity into a brief period before the end of the life span, with the expensive burdens of chronic illness occurring only in the last five to seven years of an average life of 85 years.[45]

There is significant evidence of a steady, measurable squaring of the mortality curve, showing that more and more people are living to an older age in relatively good health. The evidence also shows that medical care has little to do with the "squaring," indicating that further gains are likely to come from disease prevention and promotion of health. Fries argues that this pace can be accelerated, in effect postponing cancer and other debilitating chronic conditions until after we are dead. The effects of this postponement are shown in Figure 4–5. If evidence continues to mount in support of Fries's proposition that preventive approaches are the most important elements for postponing infirmity, the implications for the health care field are potentially profound.[46]

Resource allocations for health care can then be expected to move in this direction, from 95 percent for health care and 5 percent for prevention to perhaps 80 percent and 20 percent, respectively.

Fries estimates that under the most favorable assumptions a 15 percent decrease in health care costs could be achieved in the next twenty-five years from the compression of morbidity alone. Other reductions would depend on whether society will eliminate expensive, unnecessary technological interventions at a greater rate than it introduces new interventions. There is some evidence that the acceleration of health care costs is starting to moderate; medical inflation, which has been equal to 200 percent of the CPI for years, dropped to 150 percent of 1983's CPI. The compression of morbidity, with vaccines and cures for major diseases and "soft" but effective behavioral therapies, and changes in the health care delivery system provide the potential for decreasing the percentage of GNP consumed by health care from the current 11 percent to perhaps 6–8 percent by 2010.[47]

Health promoting programs that could lead to the compression of morbidity can be fundable through their own savings. Fries argues that promotion of health in the workplace is likely to follow an economic curve: Each dollar spent on health promotion will have a return of up to ten or twenty dollars, with the first dollars spent providing the greatest return and successive dollars providing a decreasing return until the top of the curve, estimated at approximately $200 per employee. The health promotion programs may repay employers in two years, while self-management programs (reducing doctor visits, for example) have an immediate payback. Corporation managers are likely first to redesign benefit plans and then, when that doesn't prove effective enough in reducing costs, move toward attempting to lower health costs throughout the community. Only after those attempts will they turn to the promotion of health. Corporate delay, with a workers' resistance to donating their own time to health promotion programs, may mean the marketplace will be slow to adopt the movement.

The issue remains whether or not the poorest of society will ever achieve the gains from a market-driven approach to the promotion of health. Some argue that government has a responsibility to accelerate these programs and extend them throughout society, particularly to those who have been forced out of the workplace. Government sponsorship of the promotion of health might be justified on moral grounds and on economic grounds if the programs would produce savings in programs such as Medicaid and Medicare. Others hold that it will be both more efficient and appropriate to let the market determine the extent and rapidity of health promoting programs. (See the discussion of implications in Chapter 5 for consideration of how corporate and community health promotion might evolve.)

Support for promotion of health must be considered in the context of its critics and the enormous resistance of the existing health care industry to significant movement away from the curative therapeutic model, which has been prevalent for so long. Some signs, such as the 1979 Surgeon General's Report, *Healthy People*, indicate the beginning of a slow but discernible shift to health promotion.[48]

A variety of critics disagree with Fries on the prospects for a significant compressing of morbidity. A number of epidemiologists and demographic statisticians argue against Fries's interpretation of the squaring survival curve.[49-51] The extent to which rectangularization is occurring and will occur is controversial. Even more important, the issue of what age represents the average human life span has large implications for the future of the health care system. Critics argue that if the great proportion of people do not die near the age of 85, as Fries projects, there will be far more very old people who will demand more medical services. Diseases such as dementia dramatically increase in incidence among those over 85, affecting as much as one-third of this population. Those who argue that life expectancy will keep increasing to 90 or more in coming decades look for an increasing proportion of the population budened with chronic illnesses. They argue that morbidity will not be compressed but instead extended to older and older ages, causing demand for more and more health care resources. The counterargument is that morbidity will be compressed for those living longer, thereby reducing total health care costs regardless of the lengthening of life. Likewise, some of the less preventable chronic diseases that have killed people in the past are now treatable; thus people can be kept alive, often for a long period of time. Health costs apart, some argue that the health gains involved in the compression of morbidity would lead to higher federal expenditures because of the prolonged social security and other payments that would be associated with the elimination of various chronic diseases and accidents, as shown in Table 4-1.

Table 4–1. The Potential Increases in 1983 Federal Costs If Certain Causes of Death Were Eliminated

	% Increase in Life Expectancy at Age 65	*Full Federal Cost (in billions of 1983 dollars)*
Heart disease	30.9%	$67.4
Neoplasms	8.5%	18.5
Cerebrovascular disease	7.3%	15.9
Motor vehicle and other accidents	1.2%	2.6

Based on estimates of gains in life expectancy published in National Center for Health Statistics (U.S. Public Health Service), "U.S. Life Tables by Cause of Death: 1969–71," by T.N.E. Greville, U.S. Decennial Life Tables for 1969–71, vol. 1, no. 5, 1976.

Source: Barbara Boyle Torrey and Douglas Norwood, "Death and Taxes: The Fiscal Implications of Future Reductions in Mortality."

Nutrition and Health

Of the many components comprising health and the promotion of health, increasing attention has been focused on nutrition and health. Among the public, this is reflected in the great interest in a wide variety of nutrition and diet books and programs. Nutritional assessment of the general public, as well as those individuals at risk, and the establishment of reasonable dietary guidelines will become increasingly important in the future.

Of particular interest and discussion among members of the nutrition research community are issues such as those related to the links between diet and hypertension, heart disease, and cancer. Initial studies found a strong link between sodium consumption and high blood pressure, and this implication led many consumers to decrease their sodium intake. Recent evidence suggests deficient amounts of calcium and potassium may also increase the risk of hypertension.[52] The role of nutrition in hypertension management will be a key concern in the future.

Other dietary factors affecting cardiovascular disease include polyunsaturated fats. Increased vegetable and fruit consumption, combined with a reduction in meat and poultry, may reduce blood pressure, stroke, and heart attack. High dietary intake of saturated fats has been shown to increase the low-density lipoproteins that are responsible for atherosclerosis; conversely, intake of polyunsaturated fats has been shown to increase levels of high-density lipoproteins, which play a protective role in the pathogenesis of heart disease. Moderate alcohol consumption and regular exercise have been demonstrated to increase the levels of high-density lipoproteins as well. The relationship between diet and cardiovascular disorders is one which is being continually researched.

A National Research Council report for the National Cancer Institute drew the following conclusions after a thorough examination of the scientific evidence concerning cancer and nutrition:[53]

- Many epidemiological studies show a strong association between fat intakes and cancers of the breast and the large bowels.
- High protein consumption may be associated with an increased risk of some cancers.
- Vitamin A intake is inversely related to cancers, especially lung, urinary bladder, and larynx cancers.
- Limited evidence shows intake of foods rich in vitamin C can inhibit the formation of cancer of the stomach and esophagus.
- Alcohol may act to promote liver cancer, and, in a synergistic relationship with cigarette smoke, to cause cancer of the mouth, larynx, esophagus, and respiratory tract.
- No firm conclusions can be made, given the paucity or contradictory nature of evidence, about the association between carbohydrate consumption, cholesterol, vitamin E, B-vitamins, iron, copper, zinc, molybdenum, iodine, arsenic, cadmium, and lead and their individual roles in cancer.

Recent other studies have pointed to total caloric intake as linked with incidence of cancer. Obesity also contributes to the etiology of other chronic diseases such as heart disease, hypertension, and diabetes. Organizations such as the NCI, the American Cancer Society, and other major cancer research groups are focusing greater emphasis on the diet-cancer relationship. The diet-related public health issues may change over the next few decades as new areas replace some of the current emphasis on selected chronic diseases. In any case, making the proper food choices and dietary decisions rests with the consumer. Educating and guiding the consumer to make these choices is of primary importance; it means that useful information on dietary recommendations must be provided.

Increasingly, the worksite has been effective in this education process. Corporations have begun to play a role in the health of their employees through the formation of preventative health promotion programs. Many of these programs cover nutrition education, diet, and weight loss, and some programs have nutritionists on staff to counsel and formulate individualized diets for those with specific physiological problems. Identifying those individuals who are at high risk for disease will be an integral part of the intervention process. Once these risk factors are identified, both dietary manipulation and other lifestyle changes can occur. For example, disulfites have been found to have potentially harmful effects on asthmatics; individuals with asthma must be educated on proper dietary choices, and food labeling must include the addition of disulfites.

Nutrition education will play a large part in the increased recognition of the role of diet and health. The worksite seems a logical choice to disseminate nutrition information since many employees consume two out of their three daily meals in company cafeterias. Nutrition classes can help build awareness in reading food labels, understanding the diet-disease link, and dispelling the various myths which abound about foods, weight-loss diets, vitamins, and so on. Combining nutrition education with other health components—fitness, smoking cessation, employee-assistance counseling, and occupational safety—will help employees build a more positive approach to a healthy lifestyle.

Advances in science, especially in the areas of biomedicine, food development, and food supply will have effects on the area of nutrition and health. An ever-increasing number of convenience foods are available on the market today. The boom in the frozen-food market, especially complete meals, is an example. These foods come in convenient packaging and can be prepared quickly, often in a microwave oven. Also increasing in number are formulated foods, foods such as breakfast bars that can be considered a complete meal. Another trend in food development is low-calorie food items. These low-calorie items often include complete calorie and nutrition labeling, and the exact serving size takes the guesswork out for the consumer.

Continued development of low-sodium, low-fat foods remains uncertain in the future. Their development depends on the level of sales, which in turn is determined by the influence that food labeling has had on the consumer. If nutrition education programs are successful in building awareness of the benefits of these low-sodium, low-fat foods, sales of these products can be expected to remain at a level that encourages manufacturers to continue their production.

Other food changes affecting nutrition and health include the bioengineering of new foods. Technologies are being developed that will improve agricultural products to assure better crop materials. For instance, cattle ranchers are beginning to decrease the amount of fat on the carcasses of their meat animals, which will enable the consumer to purchase leaner meat. Ranchers are also working to have cattle leave feed lots sooner, again creating leaner meat and decreasing the cost of the meat to the consumer.

Finally, many consumers are concerned about the prevalence of ostensibly nontoxic food additives, including food dyes, preservatives, and artificial flavorings. Some items generally recognized as safe by the federal government may in fact be carcinogenic. Food contamination remains a significant problem. Bacterial and viral food poisoning as a result of contamination strikes many consumers each year.

In speculating about how our knowledge of nutrition will evolve, it is likely that we will come to understand the diet-disease relationship, as well as the diet-mood relationship. Consumers are likely to take a more proactive role in the management of their health, making proper dietary choices based on sound nutrition. Dietary factors will increasingly be part of both health promotion and treatment.

Psychoneuroimmunology—The Role of Mind in Health and Healing: Beliefs, Attitudes, and Stress

An emerging notion in the reconceptualization of health and health care has been the recognition of the importance of the mind in shaping health, preventing disease, and accelerating recovery. Although much of medical practice remains wedded to techniques and procedures excessively Cartesian in their origin, there is little dispute today that the mind plays a much larger role in health and healing than has been assumed in recent decades. Many biomedical researchers are now beginning to take questions of mind-body interrelationships in the healing process much more seriously.

Leading treatment centers in the fields of cancer[54,55] and in chronic pain[56] are making much greater use of patients' mental images in diagnosis or as an adjunct to therapy. Norman Cousins' books have popularized the issue,[57] and a new organization, The Institute for the Advancement of Health, and its journal, *Advances*,[58] have recently been established to focus research efforts on these questions. Another journal on this topic, *Investigations*,[59] has been started by the Institute for Noetic Sciences, which also publishes *The Institute of Noetic Sciences Newsletter* for discussions of research and conferences on the interrelationships among emotions, the endocrine system, and the immune system. Psychoneuroimmunology (PNI) is a major focus of research on the mind-body link, exploring how the mind affects the immune system and makes us more or less susceptible to disease.[60]

Much of the future of medicine will rest on a more complete understanding of the role of the mind and its elicitation in the treatment and prevention of disease. Herbert Benson, the Harvard physician who first studied the physiological effects of the mental practices involved in transcendental meditation, now argues that the greater the relation of the practice to the person's core values, the more profound the impact.[61] For many people religion is associated with their core values. Faith healing, something which has always occurred, has grown in visibility with the rise of the charismatic movement. Major denominational churches such as the Lutherans and Episcopalians have begun to explore its significance and the role they and their members can play in health and healing.[62]

To the extent that core values are associated with religious or spiritual practices, and that these values grow, will the power involved in "faith healing" become more common? The "laying on of hands," for example, is now being studied through a federal grant because therapeutic touch has been observed to relieve pain, reduce anxiety, increase the amount of oxygen-carrying hemoglobin in the blood, and change brain waves.[63] The removal of the reductionist paradigm leaves open a variety of such questions about the spirit-mind dimension and its role by the early part of the twenty-first century.

A more familiar aspect of the mind-body link for many is stress. Yet, until recently, it would have been difficult to find a credible seminar, a thoughtful article, or research evidence supporting the critical role that stress plays in human health. Of course, stress, as it is becoming understood, is a mediating function between the mind and the body. While there is no question about the trend toward placing stress into the health and healing equation, a great deal of controversy remains over just what it is and how it can be mitigated or managed. There is evidence linking cholesterol levels in the blood with levels of stress in the individuals

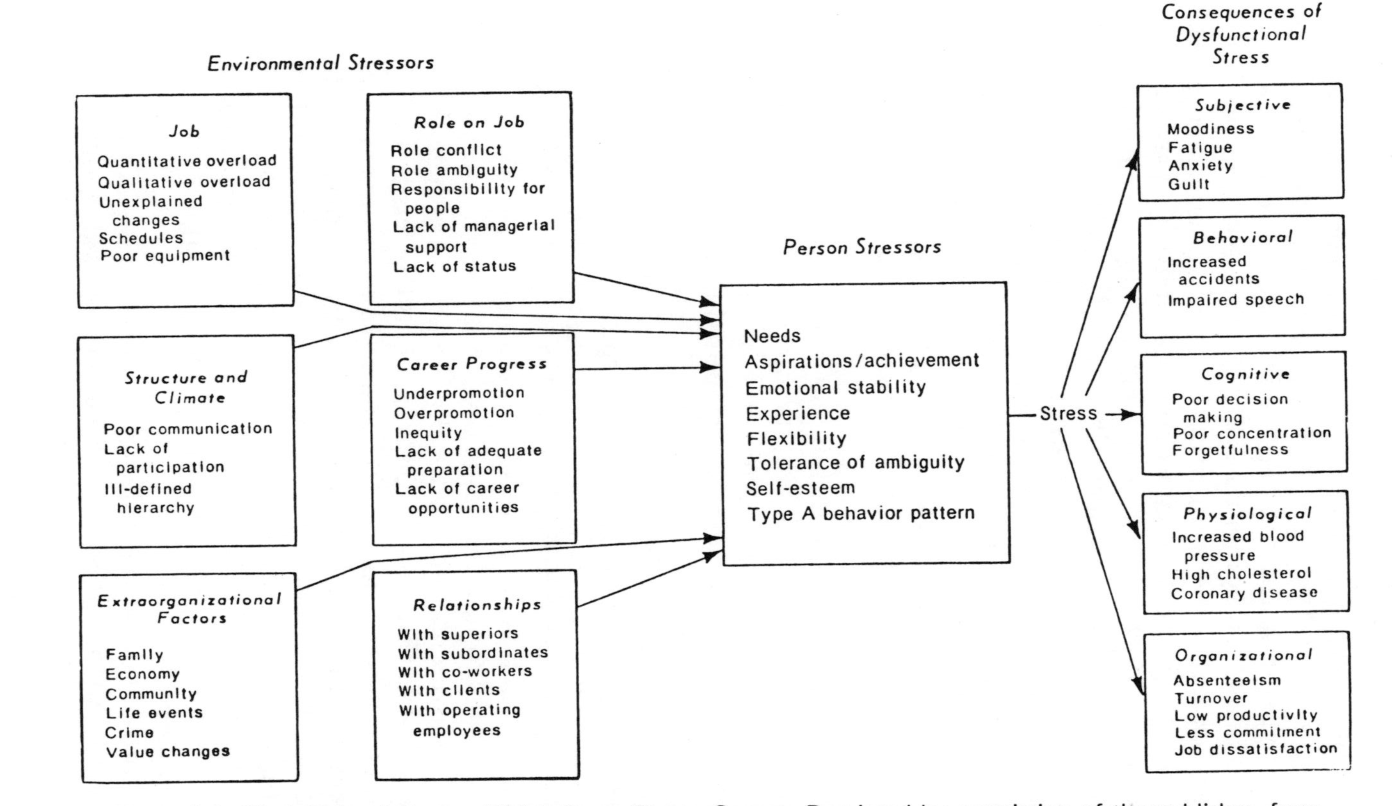

Figure 4-6 Work-Related Factors Which Cause Stress. Source: Reprinted by permission of the publisher, from "Optimizing Human Resources: A Case for Preventive Health and Stress Management," by John M. Ivancevich and Michael T. Matteson, *Organizational Dynamics*, Autumn 1980, p. 17.

studied. Moreover, emotional states associated with the impact of stress have been shown to affect recovery rates and seem as well to have a prophylactic effect for some. Stressful job demands and a lack of pride in work can contribute to heart disease and other illnesses.[64] Conversely, stress in the individual reduces productivity but can be dealt with in the work setting.[65] Blue-collar workers are often highly stressed, to the point where Arthur Shostak has argued that "blue-collar workers are a strategic element in a shared destiny (for the U.S. economy), and only as negative stress in their lives is substantially relieved does the nation's collective destiny beckon with authentic appeal."[66] Finally, Meyer Friedman's recently reported research on the prognosis for heart attack victims shows that those who adopt stress-control techniques are less likely to have recurring heart disease than those who do not.

Stress-management techniques are used increasingly, and, as Figure 4–6 indicates, the variety of potential work-related factors that cause stress and the wide range of potential dysfunctions is likely to continue to be the bridging rubric between the traditional practice of medicine and much of the work being done outside the mainstream of modern medicine. The concept, as it is currently understood, has enough "hardness" in it to be acceptable to traditional researchers and practitioners, and enough "softness" in it to be acceptable to the less conventional.

A related topic is increased stress that often accompanies unemployment, and specifically the associated health effects. Harvey Brenner, at Johns Hopkins University, has examined the recession of 1973 and 1974 to estimate the increase in mortality and crime incidence associated with various aspects of the recession. The results are shown in Table 4–2. Brenner notes that these are the direct effects of the recession. The indirect effects, including increased alcohol and cigarette use and divorce, he argues, may be even larger.[67]

Not only does unemployment sometimes lead to alcoholism, but job stress often does as well. And alcohol is only

Table 4–2. Estimated Impact of a 14.3 Percent Increase in Unemployment Rate, 3 Percent Decline in Trend per Capita Income, 5.5 Percent Increase in Business Failure Rate, 200 Percent Increase in Annual Change in Business Failure Rate (All actual changes between 1973 and 1974[3])

	Increase in stress incidence related to									
	Rise in unemployment rate		*Fall in real income trend*		*Rise in business failure rate*		*Rise in annual change in business failure rate*		*Rise in ratio of unemployment rate of males in age group 16–24 to total unemployment rate*	
Social stress indicator	*Number*	*Percent*	*Number*	*Percent*	*Number*	*Percent*	*Number*	*Percent*	*Number*	*Percent*
Total mortality	45,936	(2.3)	59,996	(3.0)	2,682	(0.1)				
Cardiovascular mortality	28,510	(2.8)	45,189	(4.4)		95,660	(9.0)			
Cirrhosis mortality	430	(1.4)	806	(2.7)						
Suicide	270	(0.98)	320	(1.1)						
Population in mental hospitals	8,416	(6.0)								
Total arrests	577,477	(6.0)								
Arrests for fraud and embezzlement	11,552	(4.8)								
Assaults reported to police	7,035	(1.1)								
Homicide[3]									403	(1.7)

[1]Direct effects only; estimates of indirect effects are discussed in Chapter V.

[2]Equations based on the years 1950–1980.

[3]Homicide figures refer to change of 9 percent in the youth unemployment ratio between 1978 and 1979.

Note: Figures in parenthesis indicate percent of total stress incidence.

Source: M. Harvey Brenner, *Estimating the Effects of Economic Change on National Well-Being*, Subcommittee on Economic Goals and Intergovernmental Policy, Joint Economic Committee of Congress (Washington, D.C.: GPO), p. 16.

one of several substances regularly used and abused to mollify office conditions. Between now and 2010, a host of new problems, opportunities, and challenges are likely to arise as research grows in the areas of psychoneuroimmunology and other aspects of brain functioning. Some argue that we will be able to mimic virtually any mental state and that it will be relatively easy to produce the neurotransmitters involved.[68]

Currently, caffeine in coffee is used by most adults as a stimulant. Cocaine is widely used for recreation and to facilitate work performance. One source said, "Each day an estimated 5,000 people, most from 18 to 34 years old, try cocaine for the first time. There are perhaps 5 million regular users."[69]

Furthermore, the boundary between appropriate and acceptable chemical enhancement of mental functioning and "illicit" use of drugs will become less distinct.[70] Some have argued that critical performance periods—for example, when taking aptitude tests for college or medical school entrance—will be a focus for chemical enhancement capability. This has led one reviewer to ask if the equivalent of the urinalysis test for steroids for athletes will be developed to prevent some from gaining "unfair" advantage in mentally competitive situations.[71] An example of the complexity of the emerging chemical enhancement of function is the use of mood-affecting drugs such as "Adam" (or MDMA, methylene dioxymethamphetamine), to encourage a greater sense of warmheartedness and empathy. In July 1985 the federal government prohibited its use.

A better idea of the complexity of the function of the body and the brain will also yield more knowledge on the full range of effects of such drugs. Related to the use of drugs to relieve stress or enhance mental performance is the stress of possible cataclysmic events and the potential for using drugs as an escape from these nightmarish specters. Generally, this use does not occupy much of our attention, although some have said that in recent years the bleak potential of nuclear war has had a significant effect on many, particularly young people.

The Rise of the Nonspecific Factors in Health and Healing

When Gregory Bateson, the famed anthropologist and writer, was dying a few years ago, he insisted on having his pet cat with him in the hospital. Of course, the hospital authorities refused him permission, and so while he was in the hospital he pretended the cat was there, and insisted that friends, family, visitors, and staff join him in the pretense. Not everyone would want a cat or a pet in the hospital, but it is doubtful that many would argue against the notion that for Bateson the cat improved his health and made his dying days more enjoyable. The issue of the cat represents a transition between an old medicine in which cats aren't important and the emerging medicine in which they are. In fact, research on heart attack victims indicates that having a pet is one of the strongest predictors of sustained recovery.[72]

A central premise of the practice of conventional medicine is that health is the result of a specific intervention focused on a specific problem. There is, on the contrary, a powerful trend toward elevating the role played by nonspecific factors—for example, social isolation, community and social activity, friends and family, and pets. It is now recognized that adverse health events and even an increased death rate can be associated with social isolation.[73]

While research on subjects like these is sometimes difficult to undertake, there are some pioneers examining nonspecific factors in health. Len Syme at Berkeley and Lisa Berkman at Yale have explored the role of social mobility, and John McKnight at Northwestern and Lowell Levin at Yale have researched the role of community and self-help groups in health.

Changing Approaches to Mental Health

Given the trends discussed thus far, mental health may be looked on differently in the years ahead. A wider variety of forces will be viewed for causal factors and for the treatment of mental health problems. The physiological basis of mental illness, for example, is becoming apparent in many research studies. Factors such as heredity and prenatal or birth trauma are being linked to mental illness, while statistical correlations show the link between economic and social factors and mental illness. Thus while the psychological and social factors of mental health are being recognized, so too are the biological components being understood. The neurotransmitter research mentioned earlier may yield very effective therapies. Simultaneously the positive or health-promoting side of mental health is gaining attention. Norman Cousins has argued that for a long time we have known of the ill effects of negative emotions but we are only beginning to explore the beneficial effects of positive emotions and actions such as love and laughter.[74] Mental wellness will increasingly become a focus of the mental health profession.

Corporate Activism in Health Care

An important set of trends in health care is the increased activism of consumers, both individual consumers and employers who, through health insurance benefits for their employees, are among the largest purchasers of health care in the United States. For decades, employers, both large and small, have played a passive role in the provision and financing of medical care services. They have been content to provide minimal infirmary and nursing services to their employees for on-the-job injuries: if they were large enough to retain a medical director, he or she supervised the infirmary and nursing services and monitored occupational and safety programs. Otherwise, employers have just payed the bill for medical care.

Many employers, if not most, are evolving from passively involved to concerned, if not proactively involved, in reducing health care costs. The impetus behind the change has been the inexorable rise of medical care costs and, to some perceptive employers, the seeming lack of relationship between costs and effectiveness. Today, for an increasing number of employers, issues of health and medical care are rapidly moving up on the corporate agenda. Workplace health and safety issues remain pressing matters for many employers, particularly as more and more worksite risks are discovered, principally in the area of toxicity. Cost containment, better put "cost management," to use Willis Goldbeck's term, has launched many employers into a wide variety of health-related activities. Chief among these are sponsorship of health-promotion, wellness, and fitness programs and participation in local and regional "business coalitions." These developments are extensively covered in Ken Pelletier's book, *Unhealthy People, Unhealthy Places*, and in publications of the Washington Business Group on Health, particularly *Business and Health* and *Corporate Commentary*.[75-77]

The increasing importance of health and health care issues to corporations has caused some companies to elevate those issues within the company and assign their management to vice presidents for health affairs. As Gilbert Collings, former chief medical officer for New York Bell System, discovered in a survey of his colleagues at twenty large American companies, 90 percent of most companies' health costs are not related to the job. "Classical occupations medicine is only dealing with 10 percent of the problem." [78] Collings argues that the distinctions between occupational health issues and general health issues within companies will blur as more companies recognize they cannot effectively be separated.

This changing role for the corporation with regard to health and health care issues is linked to emerging trends in the workplace focused on human resource development, emerging concepts of "human capital," and workplace democracy and employee-participation programs described in Chapter 4.

Consumer Monitoring of Health Care Delivery

Consumers are becoming increasingly more conscious about their care purchases. A major factor is a change in incentives and cost. Larger numbers of individuals and families no longer have first-dollar coverage; patient coverage for the first few trips to the doctor, or a percentage or charge for all care, is increasingly common. Dissatisfaction with physicians and health care generally also play a part in the shift of consumers as more assertive and discerning buyers of care, and this shift is being encouraged by employers.

For some time local consumer groups in certain cities have monitored and rated local physicians and HMO's (for example, the *Washington Consumer Checkbook* in the D.C. area). Recently a national organization, The People's Medical Society (PMS), has been created to organize consumers across the country. PMS members are rating their physicians after each visit and sending the ratings into the national office. The rating questions include whether the physician spelled out why he or she was recommending a particular treatment and what its side effects might be, as well as how the patient could prevent recurrences of the condition. Consumers also note on the form whether or not they would recommend their physician to other PMS members. PMS is also beginning to rate hospitals.[79]

Given the trends toward greater telematics in health care and the development of effective measurements of "health outcomes" for various treatments (see the following section), consumer groups such as the PMS at the national level and various local groups will become much more sophisticated in the years ahead. Electronic mail and local electronic bulletin boards will reduce the cost of taking part in such provider-rating activity. Likewise the use of home body-function monitoring, and the development of biochemically unique measures of health for individuals, will make the rating of specific providers and types of therapies very complex. Yet in the information era, consumer groups will be armed with the analytical software that insurance companies and employers will be using to ensure the quality and effectiveness of the care they buy.[80]

Measurement of Health Outcomes

As odd as it may sound, until the last ten years or so, virtually the only outcomes of patient care that were systematically measured were the levels of patient satisfaction, postoperative infection rates, and malpractice suits. In the last decade, however, particularly as the concern for the cost effectiveness of a wide variety of medical care procedures has risen, the "technology" of measuring the outcomes of patient care has been developing. In part this reflects the growing use of telematics in medicine.

In the years ahead, large-scale projects such as ARAMIS, which gathers sophisticated measures of care for arthritis patients at sixteen teaching hospitals, will have a significant impact on medical practice, particularly when combined with other developing data bases. Such measures will provide physicians with a new means to compare the results of different therapies, and it may provide consumers and payers with the ability to compare the results achieved by different practitioners.

Health outcome measures are designed to reveal the effect of health care on disability, discomfort, general economic loss (the cost of treatment plus time lost from work), iatrogenic effects, and death.[81] Their development depends on long-term collection of standardized data from large numbers of patient records. Until recent years, outcome measures were slow in developing because the traditional methods of isolating variables in double-blind experiments could not realistically be used. Researchers have since turned to survey techniques and epidemiological review of patient records to analyze large numbers of variables.

The further development of outcome measures may affect both care and reimbursement for care. There are obvious applications in the postmarket surveillance of pharmaceuticals, giving greater knowledge of the side effects and trade-offs of drugs. Less obvious, but perhaps more important, outcome measures will add to medicine's capacity to recognize the widely varying response individuals can have to medical interventions. Another development may be the use of information from health outcome measures for cost containment; payers may choose to tie reimbursement to the demonstrated effects of therapies.

Corporate groups are pursuing the development of health outcome measures under the banner of "bringing accountability to medical practice." The nature of their concern ranges from the often wide variation in the cost of similar procedures in different regions of the country, or even in the same region, to the estimate, from a Rand study, that 20 to 30 percent of all medical care is unnecessary.[82] This corporate interest will stir greater demand for the types of measures we described, as well as for reporting the data which are already collected by hospitals for their accreditation process.

Developing outcome measures is not a simple task, particularly in the arena of promoting health. There is some hope that good experimental design will allow health promotion data to be piggybacked onto the collection of other corporate data. Most large corporations, however, don't have systematic data on the number of employees and absentee days, making difficult the demonstration of such benefits as a reduction of absenteeism. Whether sufficient studies will be supported by corporate or union data gathering is unclear, and some argue that this is an area for government involvement. A similar situation exists for disease-prevention programs. Successful prevention goes unmeasured, and, paradoxically, if a program succeeds in preventing a formerly prevalent disease, funding is often reduced as a response until the disease reappears.

Notes

1. *The Economist*, "Pensions After the Year 2000," May 19, 1984, pp.59–62.
2. U.S. Department of Health and Human Services, *Health United States and Prevention Profile*, DHHS Publication No. (PHS) 84–1232 (Washington, D.C.: GPO, December, 1983).
3. Clement Bezold, "The Uncertain Future: Alternative Futures for the U.S. and Health Care," in *Pharmacy in the 21st Century*, Clement Bezold, Jerome Halperin, Richard A. Ashbaugh, Howard Binkley, eds. (Virginia: Institute for Alternative Futures and Project HOPE, 1985).
4. James F. Fries, "Health Trac," 701 Welch Road, Ste. 214, Palo Alto, California 94304.
5. Jonathan E. Fielding and Leslie M. Alexandre, "Models for Assessing Health," *Business and Health*, March 1984, pp. 5–12.
6. Phillip L. Polakoff, "Health Care in the 21st Century," *Occupational Health Safety*, September 1982, pp. 31–32, 34.
7. Clement Bezold, "Health Care in the U.S.: Four Alternative Futures," *The Futurist*, August 1982, pp. 14–19.
8. James S. Turner, "Computers, Consumers, and Pharmaceuticals," in *Pharmaceuticals in the Year 2000*, Clement Bezold (Alexandria, Va.: Institute for Alternative Futures, 1983).
9. Clement Bezold, "Medical Megatrends Reshaping Delivery and Evaluation of Care," *Modern Healthcare*, July 1984, pp. 165–167.
10. Lewis Thomas, *Lives of a Cell: Notes of a Biology Watcher* (New York: Bantam Books, 1974).
11. Arthur C. Clarke, *Report on Planet Three* (New York: Signet), pp. 129–130.
12. Michel Salomon, *Future Life* (New York: Macmillan, 1983), p.87.
13. *Ibid.*, pp. 34, 53, 99.
14. *Ibid.*, p. 82.
15. *Ibid.*, p. 44.
16. Louis F. Reichardt, "Immunologic Approaches to the Nervous System," *Science*, September 21, 1984, p. 1298.
17. Michael Salomon.
18. Charles Panati, *Breakthroughs: Astonishing Advances in Your Lifetime in Medicine, Science, and Technology* (Boston: Houghton Mifflin, 1980), pp. 142–143.
19. Clement Bezold and Jonathan Peck, "Preparing for the 2nd Pharmaceutical Age," *Pharmaceutical Executive*, May 1984, p. 35.
20. Julie Ann Miller, "Diagnostic DNA," *Science News*, August 7, 1984, p. 105.
21. *Ibid.* p. 107.

22. William Check, "New Drugs and Drug Delivery Systems in the Year 2000," in *Pharmacy in the 21st Century*, Clement Bezold, Jerome Halperin, Richard Ashbaugh, Howard Binkley, eds. (Bethesda, Md.: American Association of Colleges of Pharmacy, 1985).
23. The Health Central System, *1985–89 The Restructuring Health Industry: Progress Through Partnerships*, March 1984, pp. 28–30.
24. *Ibid.*, p. 31.
25. *Ibid.*, p. 24.
26. *Ibid.*
27. *Ibid.*, p. 25.
28. Institute for Alternative Futures, "The Prospects for Home Health Care," 1985.
29. Center for Health Management Research, *Foresight: A Publication of the Lutheran Hospital Society of Southern California*, Fall 1984, p. 2.
30. *Ibid.*
31. Arthur C. Hastings, James Fadiman, and James S. Gordon, eds., *Health for the Whole Person: The Complete Guide to Holistic Medicine* (Boulder: Westview, 1980).
32. Robin Munro, "Medicine from beyond the Fringe," *New Scientist*, January 1983, pp. 151–154.
33. Raymond F. Rosenthal and James S. Gordon, *New Directions in Medicine: A Director of Learning Opportunities* (Washington, D.C.: Aurora Associates, Inc., 1984).
34. Carol Cain, "Healers Foresee Insurance Coverage on Non-Traditional Medical Services," *Modern Healthcare*, July 1984, pp. 108–116.
35. Lori B. Andrews, *Deregulating Doctoring: Do Medical Licensing Laws Meet Today's Health Care Needs?* (Emmaus, Pa.: People's Medical Society, 1983).
36. Hastings, Fadiman, and Gordon.
37. Mary Longe, "Hospital Based Health Promotion," *The Promoter*, Winter 1985, p. 1.
38. Paul Gillette, "Competition Heats up in Wellness Market," *Modern Healthcare*, March 1, l985, p. 42.
39. James Fries and L.M. Crapo, *Vitality and Aging* (San Francisco: W.H. Freeman, 1981), p. 103.
40. Kenneth Pelletier, *Unhealthy People, Unhealthy Places* (New York: Delacore Press/Seymour Lawrence, 1982).
41. Anne K. Kiefhaber and Willis B. Goldbeck, "Worksite Wellness," in *Health Care Cost Management: Private Sector Initiatives*, Peter D. Fox, Willis B. Goldbeck, and Jacob T. Spies, eds. (Ann Arbor: Health Administration Press, 1984), pp. 120–152.
42. Paul Gillette, p. 42.

43. James F. Fries, "Aging, Natural Death and the Compression of Morbidity," *New England Journal of Medicine,* July 1980, pp. 130–135.
44. James Fries and L.M. Crapo.
45. *Ibid.*, p. 8.
46. James F. Fries, "The Compression of Morbidity: Miscellaneous Comments About a Theme," *The Geronotologist* 24, 1984.
47. Clement Bezold, "The Uncertain Future."
48. U.S. Department of Health, Education, and Welfare, *Healthy People: The Surgeon General's Report on Health Promotion and Disease Prevention,* 1979, DHEW (PHS) Publication No. 79–55071 (Washington, D.C.: GPO, July 1979).
49. Jacob A. Brody, "Ha Ha Epidemiology and the Compression of Morbidity in the Aged," *Journal of Clinical Experimental Gerontology* 4 (3): 1982, pp. 227–238.
50. Kenneth G. Manton, "Changing Concepts of Morbidity and Mortality in the Elderly Population," *Millbank Memorial Fund Quarterly* 60: 1982, pp. 183–244.
51. Edward T. Schneider and Jacob A. Brody, "Aging, Natural Death, and the Compression of Morbidity: Another View," *The New England Journal of Medicine* 309 (14): 1983, pp. 854–856.
52. David A. McCarron et al., "Assessment of Nutritional Correlates of Blood Pressure," *Annals of Internal Medicine*, May 1983, pp. 715–719.
53. National Research Council, "Diet, Nutrition and Cancer: Executive Summary of the Report of the Committee on Diet, Nutrition, and Cancer," Assembly of Life Sciences, *Cancer Research*, June 1983, pp. 3018–3023.
54. Carl Simonton, *Getting Well Again: A Step-by-Step Self-Help Guide to Overcoming Cancer for Patients and Their Families* (Los Angles: J.P. Tarcher, 1978).
55. Gerald Jampolsky, *Love Is Letting Go of Fear* (New York: Bantam, 1981).
56. D.E. Bresler, *Free Yourself from Pain* (New York: Simon & Schuster, 1979).
57. Norman Cousins, *The Healing Heart: Antidotes to Pain and Helplessness* (New York: Avon, 1983).
58. *Advances*, journal of the Institute for the Advancement of Health, 16 East 53rd Street, New York, NY 10022.
59. *Investigations*, journal, Institute for Neotic Sciences, Sausalito, CA.
60. Kenneth Pelletier, "Sound Body/Sound Mind: Psychoneuroimmunology, the Missing Link," *Medical Self-Care*, Fall 1984, p. 12.

61. Herbert Benson, *Beyond the Relaxation Response* (New York: Time Books, 1984).
62. Henry L. Letterman, ed., *Health and Healing: Ministry of the Church* (Madison, Wis.: Wheat Ridge Foundation).
63. Jane E. Brody, "Laying on of Hands Gains New Respect," *New York Times*, March 26, 1985, p. Cl.
64. Paul J. Rosch, "The Health Effects of Stress," *Business and Health*, May 1984, pp. 5–8.
65. Jean F. Duff and Patricia J. Fritts, "Stress Management for the 80s," *Business and Health*, May 1984, pp. 9–12.
66. Arthur B. Shostak, Testimony to National Mental Health Association's Unemployment Commission, 1984, p. 15.
67. M. Harvey Brenner, *Estimating the Effects of Economic Change on National Well-Being*, Subcommittee on Economic Goals and Intergovernment Policy, Joint Economic Committee of Congress (Washington, D.C. GPO, 1984).
68. J. Richard Crout, "Technology and Its Implications for Health Care," in *Pharmacy in the 21st Century*, Clement Bezold, Jerome Halperin, Richard H. Ashbaugh, Howard Binkley, eds. (Bethesda, Md.: American Association of Colleges of Pharmacy, 1985).
69. Joan M. O'Connell, "Companies Are Starting to Sniff out Cocaine Users," *Business Week*, February 18, 1985, p. 37.
70. Donald M. Michael, "It's My Mind: The Coming Struggle over Establishment and Access to Mind Altering Agents for Higher Productivity and Quality of Experience" (work in progress), 1985.
71. Michael Schrage, "Soon Drugs May Make Us Smarter," *Washington Post*, February 3, 1985, pp. C–lff.
72. Russell Jaffe, "The Future of Coronary Heart Disease: Prevention, Treatment and Care," Institute for Alternative Futures Pharmaceutical Research and Development Seminar, September 1983.
73. James Fries and L. M. Crapo, p. 118.
74. Norman Cousins.
75. Kenneth Pelletier, *Unhealthy People, Unhealthy Places*.
76. *Business and Health*, magazine, published by the Washington Business Group on Health, 229½ Pennsylvania Avenue, S. E., Washington, D. C. 20003.
77. *Corporate Commentary*, magazine, published by the Washington Business Group on Health, 229½ Pennsylvania Avenue, S.E., Washington, D.C. 20003.
78. G.H. Collings, "Examining the 'Occupational' in Occupational Medicine: The German Lecture," presented to the Joint Conference on Occupational Health, New Orleans, Louisiana, November 1983.

79. Tom Ferguson, "The People's Medical Society: Finally, An Organization for Medical Consumers!" *Medical Self-Care*, Fall 1984, pp. 9–10.
80. Clement Bezold, "Medical Megatrends Reshaping Delivery and Evaluation of Care."
81. James F. Fries, P.W. Spitz, and D.Y. Young, "The Dimensions of Health Outcomes: The Health Assessment Questionnaire, Disability and Pain Scales," *J. Rhuem* 9:789-793, 1982.
82. Willis Goldbeck, "Incentives: The Key to Reform of Medical Economics," testimony to the House Ways and Means Committee, September 13, 1984.

Chapter 5

IMPLICATIONS FOR PROMOTION OF HEALTH IN THE WORKPLACE

In this chapter, we will discuss a series of implications for future programs for promotion of health in the workplace. The material presented here should be viewed as "work in progress" since findings from other research will generate additions, refinements, and improvements.

The chapter is divided into five parts: (1) the societal changes and effects of health promotion; (2) general, particularly environmental, issues in health promotion; (3) health care management and health promotion; (4) occupational health and safety; and (5) the operation of health promotion programs.

Societal Change and Societal Effects

The Role of the Individual, the Family, and Society

Changing social, community, and family structures have a large impact on the promotion of health. Thus, family and community associations will be perceived as more central, and therefore as more relevant, to the design of future workplace programs to promote health.

As lifestyle and behavior issues take more and more of their rightful part on center stage, the critical role that self-esteem plays in self-care, motivation, and behavior change becomes even clearer. Many of the streams of current research and thinking about the nature of health and much of the fascinating new research suggest that future health promotion programs can be more targeted on this otherwise illusive subject.

The Role of Government and Politics

Government should provide more guidance for reshaping benefit packages to offer incentives for disease prevention and promotion of health. As the governmental role in stimulating demand becomes clear, there will be a need to set standards for health care benefit programs generally, as the government has through the Medicare and Medicaid programs.

Many people have characterized health promotion programs almost by definition as "blaming the victim." Many of the suggestions in the literature and otherwise offered by some policymakers to link reimbursement for medical care benefits and/or coverage in benefit plans to behavioral change take the concept of blaming the victim one large step further. This trend will require monitoring by governments and employees, lest it become a way to avoid necessary safety changes in the workplace and the larger environment.

Politics, of course, will play a large role in the promotion of health. The fundamental issue is the degree to which special interests continue, singly or in combination, to impede or at least frustrate the introduction of effective disease-prevention and health promotion programs. The evidence, both from a research and a policy point of view, frequently favoring promotion of health is becoming more and more impressive. Yet among the reasons for the relatively slow implementation of prevention and promotion programs, and for the perpetuation of resource allocation patterns favoring "cure" over prevention and clinical care over the promotion of health, special interest politics is among the largest. The most obvious sources of resistance flow from the health professions, principally the health professional who has been trained to treat but not prevent diseases, *and* from the marketplace, where strong producer organizations, such as tobacco and sugar industries, strive to maintain their market position as well as a variety of government subsidies. The signs, however, seem clear, based on the trends examined in this book: Politics will hardly disappear from the scene in the health field, but much of the traditional resistance of the type mentioned here will begin to diminish.

The Role of Medical Research and Development

Some followers of medical research and development take the position that new research and technological breakthroughs, many already on the horizon, offer the opportunity in the future for more simplified, less onerous, perhaps even "quick fix" health promotion programs. For example, if nutritional science generates food supplements biochemically tailored to individuals, and existing, albeit crude, methods of stimulation of musculature are substantially refined, the "health-seeker" in the future might achieve the image of better health through, in this case, chemistry and electricity. Others question whether such "gains" would not likely be more cosmetic than enduring. As with a number of other issues, this issue presents a dialectic, with many forces in contention and, as with all dialectical struggles, differing values underlying the perspectives of those involved in the debate.

We have argued that our understanding of the role of the mind (and indeed the emotions) in health will increase, perhaps dramatically. Some disagree, however, stating that breakthroughs in brain biochemistry may well offer the opportunity for more direct, arguably more elegant, technologies to trigger healthier mental and emotional states. The image is mind alteration through chemistry rather than mind alteration through mediation.

The Role of the Workplace

If, in the future, employers are compelled to absorb the costs of medical care at levels disproportionate to their contractual responsibilities to their employees, resources for health promotion programs may dry up. Today, "cost shifting" has become a prominent feature in the financing of medical care. Whatever employers may believe about the values and effectiveness of health promotion programs, the deployment of ever greater resources for medical care, especially with a larger retirement population, will necessarily place greater pressure on budgets for the promotion of health. Conversely, to the extent that data confirm the role of promotion of health in the compression of morbidity, companies may increasingly see health-promotion programs as a way of limiting their health care costs.

Some Emerging Ethical Issues

In its relatively brief history, few ethical issues have been raised about the conduct and implementation of programs for promotion of health in the workplace, or elsewhere for that matter. With such programs becoming more ubiquitous and with the additional evidence of effectiveness, these issues are likely to arise with more force in the future. There are three kinds of ethical concerns. The first centers on the confidentiality of health-status-related data and the uses to which such data might be put with respect to hiring, firing, promotion, and job status generally. A second cluster of issues arises from concerns about the compulsory or voluntary nature of prevention and promotion programs and is clearly related to the first set of issues. If, for example, a particular health promotion program is well substantiated in terms of reduced health risk, should employees be required to participate if they are determined to be at risk?

The third set of ethical issues is related more to scope of coverage. For example, what justification is there for an employer to withhold a patently beneficial program from certain types of employees based on job status, responsibility, job title, and so forth? This concern extends to the research setting as well. Illustratively, if the research protocol calls for a randomized trial of a new health promotion benefit and the program is determined to be cost effective, among other things because it is beneficial to its participants, what are the ethical constraints on the continuation of the project if the major goal is research. Meyer Friedman, M.D., and his associates at Mt. Zion Hospital in San Francisco faced just that dilemma in the five-year study forming the basis of his latest book, *Type A Behavior and Your Heart*. According to Friedman, after approximately two years into the longitudinal study, which examined the effect of coping skills on reducing stress levels, those program participants determined to be "Type A" personalities showed these skills are directly related to reducing the likelihood of subsequent heart failure among participants having sustained at least one heart attack prior to the study. Dr. Friedman terminated the project at that point and converted it from a research program to one offered through this hospital.

General and Environmental Concerns

The Relative Risks of Environmental Conditions and Lifestyle Factors

The forces and influences on an individual's health that are subject to the control of the individual vary with social and economic status and education. The number of those forces is relatively few compared to those influences, many of them environmental, over which the individual possesses relatively little, if any, control. This observation leads to at least two implications for the future of health promotion programs in the workplace. First, the increased effectiveness of lifestyle-intervention programs must be kept in context in programs designed to optimize health status, since any comprehensive programs for promotion of health must necessarily address factors beyond the control of the individual. This will be a key debate as health promotion programs evolve in coming years. In fact, it can be argued, as some have, that the relatively minor impacts that many of these programs have had may stem as much from their failure to address central health issues largely beyond the control of the individual, such as the impact of family issues, lifestyle pressures outside of the job, financial and time management problems, and, while at work, management style and communication patterns.

A second issue demanding attention in the design of future health promotion programs for the workplace is the interplay of environmental and lifestyle factors in alteration of health status. The most obvious example in the worksite setting is the dual exposure to asbestos and cigarette smoking. But, with the clear emergence of a multifactoral model to explain health status, many other interactions among variables must also be noted. Whose responsibility is it, for example, for the health of an employee upon whom great, perhaps unreasonable demands are made and whose resistance to breakdown or disease is being compromised by other factors such as marital discord, alcohol or drug addiction, minor obesity, and so forth, if that employee succumbs to a heart attack? This issue is clearly related to other implications in this chapter because only when future management structures reflect the multifactoral nature of health and disease will such associations be fully understood and potentially integrated into comprehensive prevention and promotion programs.

Lingering Toxic Burden. Simply put, the question is whether or not the levels of toxicity in our environment, known, suspect, and unknown, are such that any gains in health status achieved through improved medical care and through strengthened disease-prevention and health promotion programs might potentially be compromised, offset, or rendered ineffective by the ill health caused by toxicity.

The Attitudes and Values of Employees and Managers

At various points we have discussed trends leading to the treatment of employees as "human capital." Frequently, this argument presupposes a change in management attitude as the cause for this changing perception. However, there is perhaps equally compelling evidence that the "new values," logically extended into the future, may cause employees to demand changes in management style to reflect their needs. Data about employee attitudes and opinions about management suggest that many employees, particularly younger ones, believe that a major reason for lower-than-expected productivity rates is a failure of management, not of employees. In this context, it is also constructive to note that by the year 2010, three "generations" of senior management will have come and gone and that managers, like employees, reflect the values of their times.

Where Care Is Provided: The Health Institutions

At the turn of the century, health care was provided primarily in the home and community. With the rise of modern medicine, the provision of care has been progressively institutionalized. Over the last few decades, and even today, the dominant institutions of health have been the doctor's office, increasingly the multispecialty clinic (with physicians having equity in both the practice and the real estate); the lab, as often as not complex multiphasic screening systems; and the hospital, more and more often the many-bedded secondary and tertiary care facility. As a result, the face medicine presents to the world is facade and facility, not a human face. At various points in this report, issues of scale, disenfranchisement, consumer dissatisfaction, the lack of "high touch" in the provision of care and related themes and trends, when combined, suggest a major shift in the sites and settings for care giving. In the future, homes, communities, and worksites will become major delivery sites for health care. Homes will be enhanced by new and exciting self-care tools, products, and information (especially in combination with the widespread shift to ambulatory care for cost-containment reasons). Communities' importance for association, support, and environmental protection will once again become a potentially powerful setting for care and health enhancement. Worksites offer ready access to millions in our society in need of care. Increasingly perceived by both employees and managers as an optimal place for health enhancement, worksites not only can reduce the costs of medical care but offer individuals more potent relationships among improved health status, productivity, and the achievement of higher levels of performance. Such qualities are likely to make the worksite one of the central institutions of health.

Health Care Management

Total Health Management

A recently completed research effort identified twenty-six different corporate functions that are related to the health of employees. Yet with few exceptions—for example, the supervision of hypertension screening programs or low back pain clinics by corporate medical departments—most of these functions are conducted by other departments within the organization. Integration of health concerns throughout the workplace may be a fundamental development in the future: Strategies would be formulated and then endorsed by all levels of management, including the top level, then implemented at all levels. In fact, a number of small management consulting firms are beginning to offer comprehensive health management skills. One of the implications of this type of shift within employer organizations might be to elevate the responsibility for health promotion programs "higher" in organizational hierarchies. Most such programs currently reside at the middle-management level, mostly in the benefits, personnel, or human resource department, generally with the active participation of the medical department.

The Scope of Health Promotion

Even strenuous advocates of health promotion recognize that most such programs are very limited in their reach.

Whether the percentage of our population participating in typical programs is 2 or 3 percent or 10 or 15 percent, it remains the case that such programs appeal to and usually engage mostly affluent, "upscale," mostly white, upper-middle-class participants. In part, of course, this is due to the mirror image reflecting the designers of such programs. Nonetheless, the growing armamentarium of health-enhancing skills, technologies, and information offers opportunities which can and should far transcend the limited demographics of current participants.

A future challenge, however, particularly with respect to worksite programs, where the pool of potential participants is much more demographically diverse, is to design programs that reach and motivate many more people. Further, if the worksite increasingly becomes a major site for the delivery of health programs, the scope of such programs must necessarily become broader. For example, the designers of health promotion programs have assumed that the volume of salt used both in cooking and as a condiment is of paramount importance. To the vast majority of people concerned with health, in fact, putting food on the table is of paramount importance. Whether salt is added is irrelevant at best, and probably frivolous.

If evidence about the compression of morbidity holds up and employees and their representatives become more aware of the argument, more demand will arise from employees for health promotion programs in the workplace. Evidence suggests that pressure is most likely to arise from middle-aged workers who perceive that the benefits of the "compression" are more proximate to them.

Health Costs

Employers and the business community can be expected to apply increasing pressure for health promotion programs that can be demonstrated to be cost effective. This pressure will affect the management of health care institutions in communities through business representation on boards and committees. It is also likely to be transmitted through business coalitions and, perhaps most directly, through active negotiations with providers over benefits, coverage, and price.

If cost shifting accelerates, the financial burden on the private sector may discourage further investments in disease-prevention and health promotion programs because less money will be available for new programs. The converse might also be true: Mechanisms to reduce or control cost shifting, such as "all-payors" systems, should encourage private-sector investment in prevention and promotion.

The record is clear that public employers have been far slower to initiate programs to promote health. Due in large measure to the relative lack of resources in the public sector, public employers are reluctant to commit the dollars to what might be perceived as "luxuries." As promotion of health becomes a more legitimate component of work, public sector programs are increasing, and in the future this discrepancy is not likely to exist.

Occupational Health and Safety

As John Naisbitt argues, the typical worker in the United States is now a clerk. Yet by far the major focus of existing safety and health regulation is those risks arising from industrial and manufacturing undertakings. Neither the law nor practice has caught up with the rapid changes now taking place at worksites. Clearly, the time is ripe to refocus those monitoring regulations. There is little indication and, with little exception, agreement that the workforce of the future will represent patterns of the past, and indeed there is every indication that current trends will accelerate. This demands a thorough reevaluation of existing occupational health and safety regulations.

A number of trends, when taken together, suggest that the workplace of the future will be smaller in scale. Yet much of the current methodology for monitoring safety and many health risks was established both practically and epidemiologically on the assumption of large, relatively stable employee populations. Hence, if the average worksite drops dramatically in size and trend lines depicting increased employee mobility persist, new methods will be required to identify and track these types of risk in the workforce. This new technology will probably include health "smart cards" to allow for more effective tracking of the health care of employees.

The Operation of Health Promotion Programs

The Evolution of Health Promotion Programs

The lack of definitive evaluation of health promotion programs deters more rapid growth of health promotion efforts in the workplace. There is a need to fund more research, particularly where experimental designs can take advantage of information already being gathered by corporations and unions.

Health promotion programs may be seen to have already undergone two or three stages of evolution before reaching the present state of the art: (1) the early employee-assistance programs; (2) the more recent single-component programs, such as exercise and fitness and/or nutritional interventions; and (3) early and relatively unsophisticated programs of risk assessment and intervention. Today, the state of the art subsumes earlier stages but focuses on more sophisticated risk-assessment methodologies and targeted interventions for groups of employees based on the identified risk. Many emerging programs seem to transcend today's state of the art and offer hints of future programs. If the trends discussed in this book prove to be reliable guides, future health promotion programs will continue to subsume the most effective of what has been and is being done. They may also go beyond to address issues identified earlier, such as organizational culture, management and communication styles, and family and community issues.

New technology falls into at least four categories: (1) self-help and self-monitoring devices, such as home fitness equipment, including treadmills, exercycles, and so forth, and pulse readers, timers, and biofeedback devices (this would also, of course, include self-help diagnostic tools, such as arm cuffs for measuring blood pressure); (2) second-generation self-help devices, potentially culminating in what was frequently referred to earlier as the "hospital-on-the-wrist," making health assessment more precise to foster self-help; (3) sophisticated health-profiling and assessment tools such as health-risk appraisal and multiphasic screening; and (4) intervention programs focused on health risks identified through screening which include new training programs that address motivation, values and attitudes about health, and such things as communication and relationship skills. Research breakthroughs in a number of areas can be expected to enhance health promotion. This is particularly true for the mechanisms associated with the immune system, which give people much greater capacity to resist disease and enhance their health.

The shift in values by both employers and employees toward the "expressive" values described in the Aspen Institute and Public Agenda Foundation reports is integrally related to programs in the workplace based on concepts of human capital. As noted in those reports, employees are increasingly seeking expression of these evolving values on the job. Among other things, employees highly prize participation in the generation of the rules and procedures that will apply to them and seek wide discretion in the performance of their work. These concepts are likely to carry over to the design and implementation of health promotion programs, as well as to the rules and guidelines by which they are conducted. Put sharply, if a health promotion program dogmatically fixes standards by which health and fitness are measured, those standards are less likely to be accepted by employees than if such standards were generated by the participants themselves.

Evolving Needs in Health Promotion Programs

Assuming the positive evaluations of the effectiveness of disease-prevention and health promotion programs, third-party reimbursement will need to be rewritten in order to promote consumer interest in promotion of health.

Health promotion programs are invariably shaped, or at least should be, by a clear identification of the peculiar health risks faced by a given workforce. As noted throughout this book, the size and nature—and, indeed, the attitudes and values—of the workforce of the future are rapidly changing. Employees' health risks today will most certainly not be those of tomorrow. For example, the "gods and clods" forecast of a workforce with a relatively small number of "elite" knowledge workers, with wide areas of discretion on the job, and a large number of support, maintenance, and service workers would have very specific health promotion needs. This type of workforce would be very different from today's and suggests a need for different health promotion programs. There is, in fact, the danger that future programs may focus on the "elite" workers, when the real productivity payoffs and health outcome payoffs clearly lie elsewhere.

Currently, few employers are legally required to continue medical care insurance coverage for retirees and dependents. As the retirement-age population continues to grow, many companies may refuse to extend that coverage to new retirees and may withdraw that coverage from those to whom it had been extended. If the practice becomes widespread, pressure from labor unions or from Congress could either compel employers to extend coverage or force them to accept an "offer they can't refuse." Even greater immediate pressure on resources would result, of course, with more support dependent on health promotion's morbidity-compressing effects and/or productivity-enhancing effects.

A trend which runs counter to this point is the increasing evidence that morbidity in older Americans might be significantly reduced through more focused disease-prevention and health promotion programs. If James Fries's argument outlined in Chapter 4 is correct, both private and public organizations will be faced with demands for accelerated disease-prevention and health promotion programs.

The changing perceptions of employers—principally, viewing workers as a source of "human capital"—raises not only a host of implications for future health promotion programs, but some stiff challenges as well. The key challenge is to fix the place of these programs in the workplace within the emerging array of training, retraining, and human resource development programs to which employers will increasingly turn to enrich the work experience of employees and to enhance their productivity. Health promotion programs may be relegated to a secondary status, if not treated with indifference, if the values and the outcomes of such programs inadequately meet the needs and expectations of both employers and employees for shaping and texturing the work experience and enhancing productivity.

Future Challenges and Changes in Health Promotion Programs

Powerful forces and trends are breaking down the division between mental and physical health. Traditionally, health promotion programs have been cast largely in terms of physiological health, although it is indisputable that stress-management programs also focus on mental and emotional well-being. Still, few health promotion programs have been presented in terms of mental-health promotion or mental well-being. There are some exceptions, most notably, programs offered by General Mills in Minneapolis and conferences sponsored by the Washington Business Group on Health. As the relationship between positive emotional states and improved health status becomes clearer, a new dimension in the design of health promotion programs will open up.

Health promotion programs will increasingly go beyond exercise, fitness, nutrition, stress management, and smoking cessation to include enhancement of organizational change. These new goals will include the communication patterns, personnel policies, environmental policies and management styles, those areas that encompass the corporate culture and can enhance choice and participation among workers.

The impact of decentralization in the workplace necessarily poses challenges to future health promotion programs. How far this decentralization will proceed is, of course, unknown. The size of a worksite implies both advantages and disadvantages for these programs. For example, a small worksite may be able to change harmful practices more quickly and easily, while a large worksite can marshal resources to organize classes and programs. Clearly the design of the program for a workforce of thirty is very different from that for a workforce of three thousand.

There is increasing evidence that the workplace of the future, absent effective safeguards, may be even more "toxic" than it is today. At the same time, breakthroughs in genetic research and engineering, along with comparable research breakthroughs in medical engineering, may provide the means to immunize the worker against, or at least mitigate, the effects of some or most of the toxic substances. There are at least two implications for health promotion programs from these developments. The first is that employers may allocate resources to the implementation of these programs to the exclusion of health promoting programs. The second implication is that the evolution of the health promotion field will necessarily shape and be shaped by the emerging biotechnologies in the future. The biotechnology "revolution" may also bring a host of self-diagnostic and self-therapeutic tools and technologies which could be integrated into effective health promotion programs.

An existing field of program activity in promotion of health most likely to undergo dramatic, if not radical, change in the future is stress management. Our understanding of stress and its effects, and, in particular, the relationship between stress and productivity, is just taking shape. And while many existing health promotion programs can effectively integrate stress-management techniques, the potential of future stress-managing technologies is great. For example, the emerging field of "psychoneuroimmunology" described in Chapter 4—which explores the effects of psychological states, attitudes and values, emotional conditions, and mental imagery and mental training, such as visualization techniques, on health status—is just beginning to be mapped. Future health promotion programs may be much more characterized by the degree and scope of stress-management programs than by today's emphasis on exercise and fitness.

The implications of a smaller workforce, of a shortened workweek, and an older population, most of whom are retired, for community programs are enormous. We have focused on programs in the workplace rather than the community. Nevertheless, health promotion programs in the workplace in the future may increasingly rely on community settings and community-sponsored programs both as sites and as sources of innovation in program development. In fact, centrally located community programs, rather than corporate-employer programs, may become the primary health promotion resources in many communities.

Chapter 6

CONCLUSION AND SUMMARY

The time between now and the first part of the twenty-first century will see significant change in the nature of work and the nature of health and health care. The foregoing chapters have summarized the key trends in each area that will affect the development of health promotion programs in the workplace during that time period.

Demographics and the Workforce

Basic forecasts of population and the workforce provide a starting place for thinking about the future of work. The Bureau of the Census has forecast that the U.S. population will grow from 236 million in 1984 to 283 million in 2010. Yet this forecast could be higher if immigration is greater than the Bureau of the Census assumed, if we live longer, or if we have more babies than anticipated—all plausible developments. The number and percentage of those who are elderly will grow.

Family structures will continue to change. The "feminization of poverty," resulting primarily from women in low-paying jobs who are responsible for dependent children, will continue to grow. There is likely to be a persistent "underclass" of minorities who are unable to make econonic gains proportional to the majority of the population. Our lifespan is likely to increase, and the degree of illness during the lifespan, however long, is likely to decrease for many as healthier lifestyles move the onset of chronic illness later in life or lessen its severity.

The labor force, as estimated by the Social Security Administration, will grow by the year 2010 to between 141 and 153 million. In the nearer term, the Bureau of Labor Statistics forecasts that between 1982 and 1995, of the 28.7 million new jobs added to the 102.8 million in 1982, roughly one-third would be professional or managerial positions such as lawyers and accountants; slightly more than one-third would be sales or clerical workers; and less than one-third would be service workers or other operatives and laborers. This good news is offset by the displacement of manufacturing jobs by lower-paying service sector jobs and the displacement of professional and management workers, such as middle managers, by automation and expert systems.

Trends Affecting Work

The rise of the service and information economy is a major structural change. The typical worker in the United States, once a factory worker, is now a clerk. Changes in employment statistics resulting from new information technologies vary from a net increase of jobs to a loss of 30 to 50 million jobs (of a workforce between 100 and 150 million). The new technologies, particularly information processing technologies, are likely to reduce the number of middle-level workers in professional and service industries, such as accountants, consultants, and bankers.

The globalization of work is likely to continue. As wages remain lower outside the United States, manufacturing will continue to move overseas—until, that is, domestic "robot factories" become cheaper than overseas production. In the meantime satellite communications means that many service jobs are being exported to lower wage settings, and this too is likely to continue.

Economic dark clouds—the prospects for economic decline—abound. These include recessions caused by a variety of bank defaults, oil-shutoff problems, or the possibility that we may be in an economic "long-wave" or Kondratieff cycle that will lead to economic decline through the rest of this century. The 1981–82 recession was the worst since the Great Depression, yet little social unrest occurred. Much of the energy of those who lost their jobs was absorbed by the informal economy, the unrecorded cash exchanges and barters of goods and services which now equal about 15 percent of the formal monetized GNP. This informal economy is growing, often as communities reappropriate care that welfare programs monetized and as families take part in community self-reliance efforts, particularly in the areas of food and housing. The need for this informal economy grows as welfare is cut back and as jobs are lost to robots and expert systems. This raises the question of whether we will develop ways to distribute income or wealth other than through jobs and traditional welfare programs.

Decentralization of work is growing, a trend that will foster more productive work, more autonomy for workers, and innovation. Decentralization, many argue, will lead to healthier communities, particularly as many seek to become more self-reliant. Much of the innovation in the economy is in fact coming as a result of entrepreneurship, particularly as risk-takers set up new businesses to meet new needs. The economy is also experiencing the related phenomenon of disintermediation, the elimination of the middle man wherever possible.

A transition in values is another major factor changing the nature of work. The search for "expressive values," which define success on the basis of inner growth rather than material wealth, seek harmony with nature, autonomy, personal freedom, and stronger bonds of friendship, is leading to changes in workers which are likely to enhance productivity, to focus on full utilization of human potential, and to generate more respect for worker and employer. Corporate cultures which offer worker responsibility, clearly defined rights, quality of product, and a sense of ownership and participation will be better able to take advantage of the capacities of their workers, particularly those of the baby-boom generation. Recognizing the employee as human capital is a growing trend toward enhancing the capacities of employees and treating that as a major asset.

Some experts feel that an important factor in the marketplace in the years ahead will be the worthiness of products. Consumers and workers will apply broader criteria for their product choices than simply cost. The level of worker satisfaction will be affected by the societal importance of the product and its production and marketing processes. Some even forecast that hiring and retention of the best employees will depend on product worthiness.

New technologies produced from fields such as biotechnology, holography, space manufacturing, bionics, superconductivity, and materials science could revolutionize several industries. Expert systems (computer programs which process information in a manner similar to experts in a particular field) that apply particularly to business programs will affect how work is done and how many workers are needed.

The workplace itself is facing a host of changes, including flexible working hours, work done at home, support networks, stress management, new recruitment, and compensation approaches. These changes are being driven by the rise of the two-wage-earner household, broadened social and self-awareness, challenges to middle management from technology, demographics, and corporate economies. The average workweek declined to 35.1 hours in 1982, although this figure includes a growing number of part-time workers. Career shifts are becoming more frequent, as is the work-at-home practice, either through spending some days each week away from the office or having one's job or business run out of the home. Because of the uncertain future for social security and the shaky state of some private pension plans, as well as the growing size of the elderly population, there will be increasing movement to privatize retirement through programs such as IRA's, as well as interest in delaying retirement on the part of many.

Workplace health and safety trends have generally been toward greater health and safety. There are lingering concerns about toxins, however, as well as the problem that as worksites become smaller it may become harder to enforce or encourage safety requirements. Regulatory agencies, geared for large worksites, will require some time to adjust to the shrinking workplace.

The future of worker organizations is uncertain. Unions, in their traditional form, are declining in membership. In many cases management is eliminating some of the conditions which argued for unions. The restructuring economy is forcing unions to protect those already on the job, allowing new workers to take lower wage rates and dividing the loyalty of the workforce.

Health Trends

The aging of the population, particularly if morbidity is not compressed, will put a growing strain on health expenditures, both from personal savings and from pension and retirement plans. In 1982, the elderly represented 11 percent of the population and accounted for 31 percent of the personal health care expenditures. Heart disease and cancer were the two leading causes of health care expenditures as well as the leading causes of death. Although the cost of health care will continue to be a visible issue, the changes we will now describe lead to a range of forecasts from 6 to 13 percent of GNP spent for health care in the first part of the next century (in 1984 it was 10.3 percent).

A major force shaping health care in the years ahead is telematics, the collection of emerging computer and information technologies. The "artificial intelligence" or "expert systems" will apply medical knowledge directly to patients and consumers in the years ahead to facilitate promotion of health and medical treatment. This equipment will also enable treatment to focus on each person as a unique individual: People will develop their own "biochemically unique" profiles—body functioning is as unique as fingerprints.

Biomedical research breakthroughs may provide definitive cures and/or prevention of many of the current major diseases, including heart disease and cancer. While some are more pessimistic about the prospects for such advances, many experts feel they are eminently possible. Some specific areas of research include immunology which is likely to provide vaccines for a variety of diseases, including hepatitis A and B, chicken pox, herpes simplex, tooth decay and peridontal disease, and many cancers. This area might even lead as far as immunotherapy to stop the diseases of aging. Research on the brain and its neurotransmitters will yield advances in the treatment of illnesses such as Alzheimer's disease and Parkinson's disease, as well as far more effective tranquilizers and analgesics and an understanding of how to naturally adjust one's mental state. Genetic engineering and basic genetic knowledge will yield both new treatments and new ways to deliver therapies. Single-gene-defect diseases such as sickle-cell anemia may be effectively treated. Monoclonal antibodies will provide the "magic bullet" drugs that can go directly to the offending cancer or other diseased cells and deliver their toxin with very few side effects. DNA probes will provide much more sensitive diagnostic instruments, particularly for certain bacterial and viral infections and drug resistance. Drugs will be delivered through a broader range of approaches, including nasal sprays, implantable pumps, and more effective controlled-release pills.

In addition to the new biomedical technologies there will be a growth in soft technologies, those skills with which most individuals can learn to enhance their health or to treat themselves. They include such techniques as visualization and imagery and lifestyle changes involving diet, exercise, and various approaches to stress management and personal development.

The organization, delivery, and financing mechanisms of the medical care system will change more in the next five to ten years than they have changed in the last fifty years. This change will be driven by a host of economic, political, technological, and consumer-choice factors. The medical care system is diversifying into new payment/delivery forms such as HMO's and PPO's, and new delivery sites such as outpatient surgery and home care. Health care providers are aggregating. Two out of ten physicians were in some form of group practice in the early 1980's. By the year 2000 it is estimated that nine out of ten physicians will be in group practice; likewise, virtually all hospitals will be in some form of chain or association. Investor-owned or for-profit health organizations are growing not only for hospitals but nursing homes, home care, primary care, urgent care, day surgery, and diagnostic centers.

There will be an excess of physicians while a host of alternative providers and alternative therapies become increasingly common. If no changes are made, we will have 248 physicians for every 100,000 people in the United States in the year 2000, up from 191 in 1980. Since the mid-1970's consumers have been visiting doctors' offices at a declining rate. Yet during the 1980's and 1990's physicians will compete with alternative providers such as homeopaths, acupuncturists, and nurse-practitioners. This will lead to a further questioning of licensure and the likely evolution of systems for credentialing health care providers and strengthening the market and consumers to ensure quality in health care delivery.

Self-care by individuals and families will have become increasingly important, particularly since the late 1970's, and will become even more so as the tools (for example, sophisticated diagnostics and expert systems) for self-care become more powerful and easier to use.

The health promotion and wellness movements, which have paralleled and fueled the growth of self-care and the rise of many of the alternative providers, are beginning to become a major focus of employers as well as part of the lifestyle of a growing percentage of individuals in the United States. Evidence of the positive effect of health promotion programs on business productivity, as well as on medical costs, is mounting. There is a more profound and controversial argument that promotion of health will prove to be the most cost-effective force in health care over the years ahead. This is the argument in favor of the compression of morbidity, wherein healthier lifestyles can delay the onset of chronic diseases, such as cancer and heart disease, yielding a longer life and more healthful vigor, along with lower medical care costs. Critics of this approach argue that the change will not take place and that more people will live longer but they will be sicker, requiring even more medical care. There is the related argument that adequate financing for living expenses, as well as medical care, will be an important determinant in the health of the elderly.

A critical aspect of health gains will come from our understanding the relationship between nutrition and health and changing eating patterns. The effects of diet on heart disease, cancer, and diabetes are becoming more clear, with groups such as the American Heart Association and the American Cancer Society encouraging diets consisting of more fresh vegetables and fruits, less fat, and less meat, particularly red meats. Changes among many consumers preceded government and research group recommendations in this area. The worksite is likely to become an important source of information, as well as of appropriate foods, to encourage health. Bioengineering of new food crops and processed food products is likely to produce more health promoting foods in the years ahead. Food additives and the harmful effects of pesticides could encourage the market to produce more additive-free foods, and research on foods will give us a fuller understanding of the therapeutic role of foods.

Psychoneuroimmunology is the emerging discipline which considers the role of the mind in the treatment and prevention of illness. Current therapeutic advances have been made in using techniques such as imagery in the treatment of cancer and chronic pain. Research is also exploring the role the mind and attitudes have on mediating the impact of stress on the body. Stress dysfunction is a major factor in the work-related illness and productivity loss.

Stress related to job loss is likely to be an ongoing factor, with negative health consequences corresponding to the economy's transitions as workers are laid off their jobs for varying lengths of time. Stress on the job is given as a major reason for the rising use of drugs in the office. Mental health conditions will be affected by the various factors affecting health, particularly the state of the economy, attitudes toward illness, promotion of health and breakthroughs in brain research.

Corporate activism in health care will continue to be a major force as companies become more sophisticated buyers of care through their employee benefits programs and through local coalitions. Occupational medicine, as we know it, is likely to be integrated into the broader rubric of health as the boundaries between it and general health care blur.

Consumer monitoring of health care will become more sophisticated, using the various telematic devices and new measures of outcomes of care, including self-care. Linked to their neighbors, groups will emerge that give consumers immediate information on the nature and quality of health care providers in their area, much as *Consumer Reports* does for other household products. These consumer-monitoring efforts will be aided by far more effective health outcome measures currently being developed at various medical centers and by health care providers and payors.

Implications for Promotion of Health

There are a host of implications for promotion of health from the trends described in this book. A useful way to provide the highlights of those is to provide an image of health promotion in the next century. Of course, speculation about anything in the first part of the twenty-first century is a gamble. There are many trends and forcing factors in conflict: as we have repeatedly pointed out, change is dialectical in nature. In addition, what appears today to be a major forcing factor or trend may diminish in importance, in some cases rapidly and entirely, over the next twenty-five years. With these conditions in mind, and with a healthy fear of being wrong, the following are presented as some of the likely key characteristics of health promotion programs for the workplace in 2010:

- The workplace will be dramatically different with smaller units of production, more diverse worksites—including the home—and a radically different type of work that will link skills in information processing and the use of new information technologies, such as sophisticated artificial intelligence.
- Wherever people work, they will continue to assert what we have termed the growing "expressive values." They include the right of participation in management decisions affecting the quality of work and a steadily growing emphasis on the quality of products.
- Demographically, the already outmoded view of the jobsite as comprised predominately of white males will disappear. It will yield to a much richer reality of the workplace, where nearly half of all employees will be women, where workers' racial and ethnic characteristics will be much more varied, and where workers of all ages will join others in taking advantage of a variety of flexible working arrangements and settings.
- Linked to these changes in the workplace will be a new perception of health promotion and wellness programs. They will be seen as central both to workers' expressions of their values and to management's recognition of the value of enriching employees' work experience as an ethical and effective way of enhancing pro-

ductivity. If this linkage is perceived and pursued by its protagonists, promotion of health will no longer merely be a "nice to have" fringe benefit; rather, it will be acknowledged as an integral part of the work experience. If this linkage is not established, the reverse could be the case.

- Fueling the expansion and furthering the impact of health promotion programs will be research and conceptual and technological advances in the health field. They will include many unforeseeable breakthroughs. Yet almost certain to be among these advances will be far more precise and powerful health assessment tools which will focus on a wide variety of human sensitivities to environmental and nutritional factors and will lead to enhanced health as well as disease prevention and cure. Further breakthroughs in the emerging fields of psychoneuroimmunology will make today's stress-management programs seem primitive. Genetic research breakthroughs will allow the development of detailed profiles of worker sensitivities and capacities. And, lastly, a little bit more of the powerful range of human intelligence and consciousness will be discovered, enabling the development of many new "soft technologies" of the mind and body.
- The dominant institutions of health—the hospital, the clinic, the laboratory, the nursing home, and so forth—will steadily shrink in influence and scope in the decades ahead. They will yield to the home, the community center, and the workplace as the dominant settings at which health is pursued.
- As these changes occur, ongoing radical changes in the structure and nature of the health industry will continue unabated. Chief among these changes will be the disappearance of a substantial number of freestanding hospitals and the emergence of a few large health care organizations. These corporations will provide care and health-enhancing programs in a wide variety of settings, leading, among other things, to a much closer

linkage between such organizations and workplace health programs. Hence, the typical employee of the future may participate in a health promotion program under common management at a local hospital, in the home—including the home information center—at a community center, and on the job. Participation will be possible through association with a large health care organization, and employees will be able to take advantage of those benefits almost anywhere in the United States.

- These developments, and others we have discussed, present a powerful challenge to existing occupational health and safety programs, and they point to the need for adaption to an emerging, dramatically different environment. The 1984 disaster in Bhopal, India, provides a reminder that, even as we move from an industrial-based economy to one of services and information, many hazards persist. Indeed, a new wave of work-related dangers may emerge from the new technologies being introduced in the workplace, requiring more, rather than less, vigilance. There is, for example, powerful evidence that toxicity is increasing in our environment, including the workplace, and that wholly new categories of disorders and disabilities may emerge in the future. Among other implications of this change will be the emergence of "total health care management" in the workplace, concerned with all types of risks and hazards, the provision of medical care, and the advancement of health.

Although the broad outlines of change are clear, the dialectical nature of change is perhaps the most dominant feature of change itself. Nevertheless, virtually everything here points to a public dramatically more interested in its health than ever before, the evolution of a rich set of new programs and approaches for enhancing health, and the recognition of the workplace as a rich setting for health promoting activities. When these factors are combined with the emergent perception of the employee as the key source of capital, the future of health promotion programs in the workplace looks optimistic.

This view of the future may be too optimistic. It assumes the best motives joined with the best information available. There are, however, some other signs of a less positive, less promising future. For example, the same behavioral signs that can be used to construct "training" programs to enhance self-care and empower individuals can also be used to construct authoritarian, coercive power to compel healthy behavior rather than to create the conditions under which it would be voluntarily sought. Similarly, breakthroughs in genetic research and screening techniques might be used to invidiously discriminate among employees, leaving the best work only to those fortunate enough to pass genetic muster.

The future, then, is indeed contradictory. Still there is powerful evidence that as we work in the future, the pursuit of health will have become ubiquitous. And health promotion will have become synonymous with productivity, performance, and with organizations that place a premium on their people.

BIBLIOGRAPHY

The Future of Work

Adams, John D. *Transforming Work.* Virginia: Miles River Press, 1984.

American Council of Life Insurance. *The Changing Work Place: Perceptions, Reality.* Trend Analysis Program, ACLI, March 1984.

______. *Forces in Motion: Identifying Potential Crises.* Trend Analysis Program, ACLI, 1983.

Andrews, Valerie. "Hazel Henderson: Will the Real Economy Please Stand Up?" *Tarrytown Letter,* January 1984, pp. 3–7.

______. "Boom or Bust: Economic Cycle or Human Sacrifice?" *Tarrytown Letter,* January 1984, pp. 8–9.

Asimov, Issac. "The Permanent Dark Age: Can We Avoid It?" *Working in the Twenty-First Century,* C. Stewart Sheppard and Donald C. Carroll, eds. New York: John Wiley & Sons, 1980, pp.1–11.

Beckman, Robert. *The Downwave: Surviving the Second Great Depression.* New York: E.P. Dutton, 1983.

Bell, Daniel. *The Coming of Post-Industrial Society: A Venture in Social Forecasting.* New York: Basic Books, 1973.

Best, Fred. "Recycling People: Work-Sharing through Flexible Life Scheduling." *The Futurist,* February 1978, pp. 5–16.

Bezold, Clement. "Lucas Aerospace: The Workers' Plan for Socially Useful Products." In *Anticipatory Democracy,* Clement Bezold, ed. New York: Random House, 1978.

Brenner, Harvey M. *Estimating the Effects of Economic Change on National Health and Social Well-Being.* Study prepared for the Joint Economic Committee, Congress of the United States. Washington: Government Printing Office, June 15, 1984.

Brooke, James. "The Pros and Cons of 'Computer Commuting': It's Still Not the Big Apple." *New York Times,* September 23, 1984, p. F15.

Brown, Arnold, and Edith Weiner. *Supermanaging*. New York: McGraw-Hill, 1984.

Business Week. "A Work Revolution in U.S. Industry." May 16, 1983, pp. 100–110.

Callenbach, Ernest. *Ecotopia,* Berkeley, Calif.: Banyon Tree, 1975.

Cetron, Marvin. "Getting Ready for the Jobs of the Future." *The Futurist*, June 1983, pp. 15–22.

Cetron, Marvin, with Marcia Appel. *Jobs of the Future.* New York: McGraw-Hill, 1984.

Cetron, Marvin, and Thomas O'Toole. *Encounters with the Future: A Forecast of Life in the 21st Century*. New York: McGraw Hill, 1984.

Clarke, Arthur C. *Profile of the Future.* New York: Holt, Rheinhart and Winston, 1984.

Cleveland, Harlan. "The Twilight of Hierarchy: Speculations on the Informatization of Society." Prepared for the Symposium on Information Technologies and Social Transformation for the National Academy of Engineering, October 4, 1984.

Coates, Joseph F. "The Changing Nature of Work." *The World of Work: Careers and the Future,* Howard F. Didsbury Jr., ed. Bethesda, Md: World Future Society, 1983, pp. 25-32.

Coates, Vary T. "The Potential Impacts of Robotics." *The Futurist*, February 1983, pp. 28–32.

Coates, Vary T. et al. *Toxics '95: The Outlook of Factors and Trends for Toxic Chemicals.* Office of Pesticides and Toxic Substances, U.S. Environmental Protection Agency, May 1984.

Cole, Sam, and Ian Miles. *Stacking Up the Chips: The Distributional Impact of New Technologies.* London: Francis Pinter, 1985.

Cooper, Mary H. "Healthcare: Pressure for Change." *Educational Research Reports.* Washington, D.C.: Congressional Quarterly, August 10, 1984.

The Cornucopia Project. *Empty Breadbasket? The Coming Challenge to America's Food Supply and What We Can Do about It.* Emmaus, Pa: Rodale, 1981.

Dentzer, Susan. "A Touch of 'Made in America.'" *Newsweek,* October 27, 1984, p. 102.

Didsbury, Howard, ed. *The World of Work: Careers and the Future.* Bethesda, Md.: World Future Society, 1983.

The Economist. "Pensions After the Year 2000," May 19, 1984, pp. 59–62.

Etzione, Amitai. *An Immodest Agenda.* New York: McGraw-Hill, 1983.

Family Service America. "The State of Families 1984–85." New York, 1984.

Farnsworth, Clyde. "The Too-Mighty Dollar Takes a Toll of Manufacturing Jobs." *New York Times,* September 23, 1984, p. E3.

Forrester, Jay W. "Managing the Next Decade in the Economy." Paper for the New Eonomy Conference Sponsored by the Joint Economic Committee and the Small Business Subcommittee on General Oversight and the Economy and the Congressional Clearinghouse on the Future, Photocopy, June 6, 1984.

Fromm, Erich. *To Have or To Be.* London: Jonathan Cape, 1978.

The Futurist. 1984, 1985 issues. Published by *The World Future Society,* 4916 St. Elmo Avenue, Bethesda, MD 20814-5089.

Ginzberg, Eli. "The Mechanization of Work." *Scientific American*, September 1982, pp. 67-75.

Goldbeck, Willis. "Incentives: The Key to Reform of Medical Economics." Testimony to the House Ways and Means Committee, September 13, 1984.

Gutman, Peter M. "The Subterranean Economy Five Years Later." *Across the Board*, February 2, 1983, pp. 24-31.

Harman, Willis. "Future of the Earth—The Role of Corporations." Paper presented at The Other Economic Summit, 42 Warriner Gardens, London, England, SW11 4DU, June 6-10, 1984.

______. "Work." In *Millennium: Glimpses into the 21st Century*, Alberto Villoldo and Ken Dychtwald, eds. Boston: Houghton Mifflin, 1981.

Hawken, Paul. *The Next Economy*. New York: Ballantine Books, 1983.

Henderson, Hazel. *The Politics of the Solar Age: Alternatives to Economics*. Garden City, N.Y.: Anchor Press/Doubleday, 1981.

Inglehart, Ronald. *The Silent Revolution*. Princeton, N.J.: Princeton University Press, 1977.

Jacobs, Jane. *Cities and the Wealth of Nations: Principles of Economic Life*. Toronto: Random House, 1984.

Jones, Barry. *Sleepers Wake! Technology and the Future of Work*. New York: Oxford University Press, 1982.

Kanter, Rosabeth Moss. *Change Masters*. New York: Simon & Schuster, 1983.

Kerr, Clark, and Jerome Rosow, eds. *Work in America*. New York: Van Nostrand and Reinhold, 1979.

Kieffer, Jarold A. "The Coming Opportunity to Work Until You're 75." *Washington Post*, September 9, 1984, p. D1.

Leontief, Wassily W. "The Distribution of Work and Income." *Scientific American*, September 1982, pp. 188 ff.

Levitan, Sar A., and Clifford M. Johnson. *Second Thoughts on Work*. Kalamazoo, Mich.: W.E. Upjohn Institute for Employment Research, 1982.

Levitt, Theodore. "The Globalization of Markets." *Harvard Business Review*, May-June 1983, pp. 92-102.

McGahey, Richard. "High Tech, Low Hopes." *The New York Times*, May 15, 1983, p. E21.
Main, Jeremy. "Work Won't Be the Same Again." *Fortune*, June 28, 1983, pp. 58–65.
Michael, Don. M. "Competence and Compassion in an Age of Uncertainty." World Future Society *Bulletin*, January/February 1983, pp. 1–6.
Michael, Donald M. *Cybernation: The Silent Conquest*. Santa Barbara, Calif.: Center for the Study of Democratic Institutions, 1962.
Mitchell, Arnold. *The Nine American Lifestyles: Who We Are and Where We're Going*. New York: Macmillan, 1983.
Morris, David. *Self-Reliant Cities: Energy and the Transformation of Urban America*. San Francisco: Sierra Club Books, 1982.
Naisbitt, John. *Megatrends*. New York: Warner, 1982.
Newsweek. "The Kindness of Strangers: America's Debt Is Financed by Foreigners and They May Soon Decide to Pull Out," February 27, 1984.
New York Stock Exchange. *U.S. International Competitiveness: Perception and Reality*. New York: NYSE, Office of Economic Research, August 1984.
O'Neill, Gerald K. *The Technology Edge*. New York: Simon & Schuster, 1983.
O'Toole, James. *Making America Work: Productivity and Responsibility*. New York: Continuum, 1981.
Perelman, Lewis J. *The Learning Enterprise: Adult Learning, Human Capital and Economic Development*. Washington, D.C.: The Council of State Planning Agencies, 1984.
Reich, Robert B. *The Next American Frontier*. New York: New York Times Books, 1983.
Riche, Richard W., Daniel E. Hecker, and John U. Burgan. "High Technology Today and Tomorrow: A Small Slice of Employment." *Monthly Labor Review*. Washington, D.C.: Bureau of Labor Statistics, November 1983, pp. 50–58.
Robertson, James. "What Comes After Full Employment?" Paper presented at The Other Economic Summit, 42 Warriner Gardens, London, England, SW11 4DU, June 6–10, 1984.
______. *The Sane Alternative*. Ironbridge, Great Britain: Robertson/Spring Cottage, 1983.
Rosow, Jerome M., and Robert Zager. "Punch out the Time Clocks." *Harvard Business Review*, March–April 1983, pp. 12–30.
Roybal, Edward R. "The Social Security/Medicare Trustees' Report," Letter to Members of the U.S. House of Representatives House Select Committee on Aging, April 6, 1984.

Scientific American. "The Mechanization of Work," November 1982 (entire issue).
Security Pacific National Bank. *Trends: Executive 1990: A Search for Leadership*. Futures Research Division, 1984.
______. *Trends: The Once and Future Economy*. Futures Research Division, 1984.
______. *Trends: Help Wanted 1990*. Futures Research Division, 1983.
Shostak, Arthur B. "Options for Blue-Collar Workers," *Business and Health*, May 1984, pp. 13–15.
Sorokin, Pitirim. *Social and Cultural Dynamics*. New York: American Book Company, 1937–1941, 4 vols.
Staines, Graham L., and Joseph H. Pleck. *The Impact of Work Schedules on the Family*. Ann Arbor, Mich.: Institute for Social Research, 1983.
"The States in 1990." *American Demographics*, December 1983, pp. 21–23, 45.
Tarrytown Letter. "Supermanaging in the 80s," pp. 3–6.
Toffler, Alvin. *The Adaptive Corporation*. New York: McGraw-Hill, 1985.
______. *The Third Wave*. New York: Bantam, 1981.
Torrey, Barbara Boyle, and Douglas Norwood. "Death & Taxes: The Fiscal Implications of Future Reductions in Mortality," xerox, 1984.
U.S. Bureau of the Census. *Projections of the Population of the United States, By Age, Sex, and Race: 1983–2080*, Series P–25, No. 952. Washington, D.C.: GPO, May 1984.
______. *Statistical Abstract of the United States: 1984*, 104th ed. Washington, D.C.: GPO, 1983.
U.S. Bureau of Labor Statistics. *Occupational Injuries and Illnesses in the United States by Industry, 1982*, Bulletin 2196. Washington, D.C.: U.S. Department of Labor, 1984.
______. *Employment Projections for 1995*, Bulletin 2197. Washington, D.C.: GPO, March 1984.
______. *Occupational Injuries and Illnesses in the United States by Industry, 1978*, Bulletin 2078. Washington, D.C.: U.S. Department of Labor, August 1980.
______. Janet L. Norwood, BLS Commissioner. Correspondence and forecasts to the year 2000. Washington, D.C.: Bureau of Labor Statistics, forecasts unpublished.
U.S. Congressional Clearinghouse on the Future. *Tomorrow's Elderly*. A report for the Select Committee on Aging, House of Representatives, Comm. Pub. No. 98–457, October 1984.
U.S. Office of Technology Assessment. *Automation and the Workplace: Selected Labor, Education, and Training Issues*. Washington, D.C.: GPO, March 1983.

U.S. Social Security Administration. *Economic Projections for OASDI Cost Estimates, 1983*, Publication Number 11-11537, Actuarial Study No. 90. Washington, D.C.: SSA, February 1984.

United Way of America. *Scenarios: A Tool for Planning in Uncertain Times*. Alexandria, Va.: United Way of America, 1984.

Walford, Roy L. *Maximum Life Span*. New York: Avon, 1984.

White House Conference on Aging. U.S. Department of Health and Human Services, June 2, 1982.

Yankelovich, Daniel et al. *Work and Human Values: An International Report on Jobs in the 1980s and 1990s*. New York: Aspen Institute for Humanistic Studies, 1983.

The Future of Health

Advances. Journal of the Institute for the Advancement of Health, 16 East 53rd Street, New York, N.Y., 10022.

American Council on Life Insurance. *Health Care: Three Reports from 2030 A.D.* Trend Analysis Program, ACLI, 1980.

Andrews, Lori B. *Deregulating Doctoring: Do Medical Licensing Laws Meet Today's Health Care Need?* Emmaus, Pa.: People's Medical Society, 1983.

Benson, Herbert. *Beyond the Relaxation Response* (New York: Times Books, 1984).

Clement Bezold, "Health Care in the U.S.: Four Alternative Futures." *The Futurist*, August 1982, pp. 14–19.

———. *The Future of Pharmaceuticals*. New York: John Wiley, 1981.

———. "Medical Megatrends Reshaping Delivery and Evaluation of Care," *Modern Healthcare*, July 1984, pp. 165–167.

———. "The Uncertain Future: Alternative Futures for the U.S. and Health Care." In *Pharmacy in the 21st Century*, Clement Bezold, Jerome Halperin, Richard A. Ashbaugh, Howard Binkley, eds. Virginia: Institute for Alternative Futures and Project HOPE, 1985.

Bezold, Clement and Jonathan Peck. "Preparing for the 2nd Pharmaceutical Age." *Pharmaceutical Executive*, May 1984, pp. 32–39.

Bowles, L.T. "The 21st Century Physician." *Journal of Medical Education*, July 1982, p. 570.

The Brain Mind Bulletin. Newsletter—coverage of literature on brain research, especially, neurotransmitters, Leading Edge, P.O. 42247, Los Angeles, CA 90042.

Brenner, Harvey M. *Estimating the Effects of Economic Change on National Well-Being*. Subcommittee on Economic Goals and Inter-governmental Policy, Joint Economic Committee of Congress, Washington: GPO, 1984.

Bresler, D.E. *Free Yourself from Pain*. New York: Simon & Schuster, 1979.

Brody, Jacob A. "Ha Ha Epidemiology and the Compression of Morbidity in the Aged." *Journal of Clinical Experimental Gerontology* 4(3): 1982, pp. 227–238.

Brody, Jane E. "Laying on of Hands Gains New Respect," *New York Times*, March 26, 1985, p. C1.

Brown, Michael S. and Joseph L. Goldstein. "How LDL Receptors Influence Cholesterol and Atherosclerosis." *Scientific American*, November 1984, pp. 58–66.

Business and Health. Magazine, published by the Washington Business Group on Health, 922 Pennsylvania Ave., S.E., Washington, D.C., 20003.

Cain, Carol. "Healers Foresee Insurance Coverage of Non-Traditional Medical Services." *Modern Healthcare*, July 1984, pp. 108–116.

Canadian Hospital Association. "Exploring the Future of Hospitals in Canada: A Definitive Study," September 1984.

Carlson, Rick J. *The End of Medicine*. New York: John Wiley, 1975.

______. ed. *The Frontiers of Science and Medicine*, London: Wildwood House, 1975.

______. "The future of health care in the United States," In *Health for the Whole Person*. Arthur Hastings et al., eds. Boulder: Westview Press, 1980.

Center for Health Management Research. *Foresight: A Publication of the Lutheran Hospital Society of Southern California*, Fall 1984.

Chandra, Ranjit Kumar. "Nutrition, Immunity, and Infection: Present Knowledge and Future Directions." *The Lancet*, March 26, 1983, pp. 688–691.

Check, William. "New Drugs and Drug Delivery Systems in the Year 2000." In *Pharmacy in the 21st Century*, Clement Bezold, Jerome Halperin, Richard H. Ashbaugh, Howard Binkley, eds. Bethesda, Md.: American Association of Colleges of Pharmacy, 1985.

Clarke, Arthur C. *Report on Planet Three*. New York: Signet.

Collings, G.H. "Examining the 'Occupational' in Occupational Medicine: The German Lecture," Presented to the Joint Conference on Occupational Health, New Orleans, Louisiana, November 1983.

Cooper, Mary H. "Healthcare: Pressure for Change." *Educational Research Reports*. Washington, D.C.: Congressional Quarterly, August 10, 1984.

Corporate Commentary. Magazine, published by the Washington Business Group on Health, 922 Pennsylvania Ave., S.E., Washington, D.C., 20003.

Cousins, Norman. *The Healing Heart: Antidotes to Pain and Helplessness*. New York, Avon, 1983.

Crout, Richard J. "Technology and Its Implications for Health Care." In *Pharmacy in the 21st Century*, Clement Bezold, Jerome Halperin, Richard H. Ashbaugh, Howard Binkley, eds. Bethesda, Md.: American Association of Colleges of Pharmacy.

Dawson, John. "Can Computers Replace Doctors?" *Private Practice*, June 1983, pp. 68–71.

DiBiaggio, J.A. "A 20-year Forecast for Academic Medical Centers." *New England Journal of Medicine*, January 1981, pp. 228–30.

Duff, Jean F. and Patricia J. Fritts. "Stress Management for the 80s." *Business and Health*, May 1984, pp. 9–12.

The Economist. "Pensions After the Year 2000." May 19, 1984, pp. 59–62.

Emery, Fred. "Public Policies for Healthy Workplaces." *Beyond Health Care*, Trevor Hancock, ed. Canadian Journal of Public Health (forthcoming).

Ferguson, Tom "The People's Medical Society: Finally, An Organization for Medical Consumers!" *Medical Self-Care*, Fall 1984, pp. 9–10.

Fielding, Jonathan E. and Leslie M. Alexandre. "Models for Assessing Health." *Business and Health*, March 1984, pp. 5–12.

Freeland, Mark, G. Calat, and C. E. Schendler. "Projections of National Health Expenditures, 1980, 1985, and 1990." *Health Care Finance Review*, Winter 1980, pp. 1–27.

Fries, James F. "The Compression of Morbidity: Miscellaneous Comments About a Theme." *The Gerontologist* 24, 1984.

———. "Aging, Natural Death and the Compression of Morbidity." *New England Journal of Medicine*, July 1980, pp. 130–135.

———. "Health Trac." 1984. 701 Welch Road, Ste. 214, Palo Alto, CA 94304.

Fries, James and L.M. Crapo. *Vitality and Aging*. San Francisco: W.H. Freeman, 1981.

Future Survey: A Guide to the Literature and Future Survey Annual, World Future Society, 4916 St. Elmo Ave., Bethesda, MD, 20814-5089.

The Futurist, magazine, World Future Society, 4916 St. Elmo Ave., Bethesda, MD, 20814-5089.

Gillette, Paul. "Competition Heats up in Wellness Market." *Modern Healthcare*, March 1, 1985, pp. 42–44.

Ginzberg, Eli. "Health Care Forecast: Adapt to Changing Environment." *Hospital Progress*, March 1982, pp. 10–12.

Goldbeck, Willis. "Bringing Accountability to Medical Practice: A Consumer Challenge." Testimony to the U.S. Senate Appropriations Committee, November 19, 1984.

______. "Future Outlook: Key to Corporate Success Lies in Past Lessons and in Worker Participation." *Business and Health*, January/February 1984, p. 55.

Goldsmith, Jeff C. *Can Hospitals Survive? The New Competitive Health Care Market*. Homewood, Ill.: Dow Jones-Irwin, 1981.

Guttmacher, Sally. "Ethics of Screening at the Work Place." *Business and Health*, March 1984, pp. 23–26.

Hammer, Signe. "The Mind as Healer." *Science Digest*, April 1984, p. 47 ff.

Hancock, Trevor. "A Quarter Century of Wellbeing and Health: The Twenty-fifth Annual Report of the Department of Wellbeing and Health to the People of Ontario 2010 A. D.," Toronto Alternative Futures Health Network, January 1983.

______. "Beyond Health Care: Creating a Healthy Future."*The Futurist*, August 1982.

Hastings, Arthur C., James Fadiman, and James S. Gordon, eds. *Health for the Whole Person: The Complete Guide to Holistic Medicine*. Boulder: Westview, 1980.

The Health Central System. *1985–89 The Restructuring Health Industry: Progress Through Partnerships*, March 1984.

Heller Research Corporation. *Health Care Practices and Perceptions: A Consumer Survey of Self-Medication*, 1984.

Illich, Ivan. *Medical Nemesis: The Expropriation of Health*. New York: Bantam Books, 1976.

Institute for Alternative Futures. "The Prospects for Home Health Care," 1985.

Investigations. Journal, Institute for Neotic Sciences, Sausalito, CA.

Ivancevich, John M. and Michael T. Matteson. "Optimizing Human Resources: A Case for Preventive Health and Stress Management." *Organization Dynamics*, Autumn 1980, pp. 5–25.

Jaffe, Russell. "The Future of Coronary Heart Disease: Prevention, Treatment and Care." Institute for Alternative Futures Pharmaceutical Research and Development Seminar, September 1983.

Gerald Jampolsky. *Love Is Letting Go of Fear*. New York: Bantam, 1981.

Joossens, J.V. and J. Geboers. "Cardiovascular Diseases, Cancer, and Nutrition" (editorial). *Acta Cardiologica* 38 (1) 1983 pp. 1–12.

Kaiser, Leeland R. "Futurism in Health Care," "Futurism—Part II," "Futurism—Part III," *Hospital Forum*, November-December 1980, pp. 26–28: March-April 1981, pp. 20–21: May-June 1981, pp. 63–66.

Kane, Robert. "The Information Revolution in Health Care—Knowing What Works and Paying for It." Institute for Alternative Futures Pharmaceutical Research and Development Seminar, 1984.

Kiefhaber, Anne K. and Willis B. Goldbeck. "Worksite Wellness," In *Health Care Cost Management: Private Sector Initiatives*, Peter D. Fox, Willis B. Goldbeck, and Jacob T. Spies, eds. Ann Arbor: Health Administration Press, 1984, pp. 120–152.

Krill, M.A. and R.R. Gayner. "What Is the Future of Prepaid Medical Practice?" *Medical Group Management*, January–February 1982, pp. 42–47.

Lerner, Michael. *Summary Report on Alternative and Adjunctive Cancer Therapies*, Bolinas, Calif.: Commonweal, June 1983.

Lesse, Stanley. *The Future of the Health Sciences: Anticipating Tomorrow*. New York: Irvington Publishers, 1981.

Letterman, Henry L. ed. *Health and Healing: Ministry of the Church*. Madison, Wis.: Wheat Ridge Foundation.

Longe, Mary. "Hospital Based Health Promotion." *The Promoter*, Winter 1985, p. 1.

McCarron, David A. et al. "Assessment of Nutritional Correlates of Blood Pressure." *Annals of Internal Medicine*, May 1983, pp. 715–719.

Maxmen, Jerrold S. *The Post-Physician Era: Medicine in the Twenty-first Century*. New York: John Wiley, 1976.

Manton, Kenneth G. "Changing Concepts of Morbidity and Mortality in the Elderly Population." *Millbank Memorial Fund Quarterly 60 1982, pp. 183–244.*

Medical Self-Care. Magazine, P.O. Box 717, Inverness, CA 94937.

Michael, Donald M. "It's *My* Mind: The Coming Struggle over Establishment and Access to Mind Altering Agents for Higher Productivity and Quality of Experience" (work in progress), 1985.

———. *Cybernation: The Silent Conquest*. Santa Barbara, Calif.: Center for Study of Democratic Institutions, 1962.

Miller, Alfred E. and Maria G. Miller. *Options for Health and Health Care.* New York: John Wiley, 1981.

Miller, Julie Ann. "Diagnostic DNA." *Science News*, August 7, 1984, pp. 104–07.

Moschetto, C.A. "Predictions About the Future of LTC." *Nursing Homes*, September–October 1981, pp. 42–53.

Munro, Robin. "Medicine from beyond the Fringe." *New Scientist*. January 1983, pp. 151–154.

Murnaghan, J.H. "Health Indicators and Information Systems for the Year 2000." *Annual Review of Public Health*, 1981, pp. 259–261.

National Research Council. "Diet Nutrition and Cancer: Executive Summary of the Report of the Committee on Diet, Nutrition, and Cancer." Assembly of Life Sciences, *Cancer Research*, June 1983, pp. 3018–3023.

O'Connell, Joan M. "Companies Are Starting to Sniff out Cocaine Users." *Business Week*, February 18, 1985, p. 37.
Panati, Charles. *Breakthroughs: Astonishing Advances in Your Lifetime in Medicine, Science, and Technology*. Boston: Houghton Mifflin, 1980.
Health, Paradigm. "Brief to Royal Commission on the Economic Union." Toronto, 1984.
Pearson, Durk and Sandy Shaw. *Life Extension*. New York: Warner Books, 1982.
Pelletier, Kenneth. "Sound Body/Sound Mind: Psychoneuroimmunology, The Missing Link." *Medical Self-Care*, Fall 1984, p. 12.
______. *Unhealthy People, Unhealthy Places*. New York: Delacorte Press/ Seymour Lawrence, 1982.
______. *Longevity: Fulfilling Our Biological Potential*. New York: Delacorte and Delta, 1981.
Polakoff, Phillip L. "Health Care in the 21st Century." *Occupational Health Safety*, September 1982, pp. 31–32, 34.
Pritikin, Nathan. *Personal communication*, November 7, 1984.
Public Services Laboratory, Georgetown University. "Cost of Disease and Illness in the United States in the Year 2000." *Public Health Reports*, 1978, pp. 493–588.
Reichardt, Louis F. "Immunologic Approaches to the Nervous System." *Science*, September 21, 1984, pp. 1294–1299.
Robinson, Pauline K. "Facing Up to Retiree Health Care Costs," *Business and Health*, June 1984, pp. 23–26.
Rosch, Paul J. "The Health Effects of Stress." *Business and Health*, May 1984, pp. 5–8.
Rosenthal, Raymond F. and James S. Gordon. *New Directions in Medicine: A Directory of Learning Opportunities*. Washington, D.C.: Aurora Associates, Inc., 1984.
Rouse, I.L. et al. "Vegetarian Diet and Blood Pressure" (letter to the editor). *The Lancet*, September 24, 1983, pp. 742–743.
Rushman, Robert F. *National Priorities for Health: Past, Present, and Projected*. New York: John Wiley, 1980.
______. *Humanizing Health Care: Alternative Futures for Medicine*. Cambridge, Mass.: MIT Press, 1975.
Salomon, Michel. *Future Life*. New York: Macmillan, 1983.
Schneider, Edward I. and Jacob A. Brody. "Aging, Natural Death, and the Compression of Morbidity: Another View." *New England Journal of Medicine 309 (14), October 16, 1983, pp. 854–856.*
Schrage, Michael. "Soon Drugs May Make Us Smarter." *Washington Post*, February 3, 1985, pp. C–1ff.
Shostak, Arthur B. Testimony to National Mental Health Association's Unemployment Commission, 1984.

Simonton, Carl. *Getting Well Again: A Step-by-Step Self-Help Guide to Overcoming Cancer for Patients and Their Families*. Los Angeles: J.P. Tarcher, 1978.

Starr, Paul. *The Social Transformation of American Medicine*. New York: Basic Books, 1982.

Teeling-Smith, George. "Epidemics of the Future." In *Medicines in the Year 2000*, George Teeling-Smith and Nicholas Wells, eds. London: Office of Health Economics, 1979.

Teeling-Smith, George, and Nicholas Wells. *Medicines for the Year 2000*. London: Office of Health Economics, 1979.

Thomas, Lewis. *Lives of a Cell: Notes of a Biology Watcher*. New York: Bantam Books, 1974.

Turner, James S. "Computers, Consumers, and Pharmaceuticals." *Pharmaceuticals in the Year 2000*. Alexandria, Va.: Institute for Alternative Futures, 1983.

U.S. Department of Agriculture. *1983 Handbook of Agricultural Charts*. Washington, D.C.: USDA, 1983.

U.S. Department of Health and Human Services. *Health United States and Prevention Profile*, DHHS Publication No. (PHS) 84–1232. Washington, D.C. GPO, December 1983.

U.S. Department of Health, Education, and Welfare. *Healthy People: The Surgeon General's Report on Health Promotion and Disease Prevention, 1979*, DHEW (PHS) Publication No. 79–55071. Washington, D.C.: GPO, July 1979.

Villoldo, Alberto and Ken Dychtwald, eds. *Millennium: Glimpses into the 21st Century*. Los Angeles: J.P. Tarcher, 1981.

Vollset, S.E. and E. Bjelke. "Does Consumption of Fruit and Vegetables Protect Against Stroke?" (letter to the editor). *The Lancet*, September 24, 1983 p. 702.

Walford, Roy L. *Maximum Life Span*. New York: Avon, 1984.

Weed, Lawrence L. "Physicians of the Future," *New England Journal of Medicine*, April 1981, pp. 903–907.

Weil, Andrew. *Health and Healing: Understanding Conventional and Alternative Medicine*. Boston: Houghton Mifflin Company.

Wells, Nicholas. *The Second Pharmacological Revolution*. London: Office of Health Economics, 1983.

Young, Eleanor A. "Evidence Relating Selected Vitamins and Minerals to Health and Disease in the Elderly Population in the United States: Introduction." *The American Journal of Clinical Nutrition*, November 1982, pp. 979–985.

INDEX